Integrated Networking, Caching, and Computing

Integrated Networking, Caching, and Computing

F. Richard Yu
Tao Huang
Yunjie Liu

CRC Press
Taylor & Francis Group
Boca Raton London New York

CRC Press is an imprint of the
Taylor & Francis Group, an **informa** business

CRC Press
Taylor & Francis Group
6000 Broken Sound Parkway NW, Suite 300
Boca Raton, FL 33487-2742

© 2018 by Taylor & Francis Group, LLC
CRC Press is an imprint of Taylor & Francis Group, an Informa business

No claim to original U.S. Government works

Printed on acid-free paper
Version Date: 20180428

International Standard Book Number-13: 978-1-138-08903-7 (Hardback)

Visit the Taylor & Francis Web site at
http://www.taylorandfrancis.com

and the CRC Press Web site at
http://www.crcpress.com

Contents

Chapter 1

Overview, Motivations and Frameworks

New mobile applications, such as natural language processing, augmented reality and face recognition [1], have emerged rapidly in recent years. However, conventional wireless networks solely focusing on communication are no longer capable of meeting the demand raised by such applications not only on high data rates, but also on high caching and computing capabilities [2]. Although the pivotal role of communication in wireless networks can never be overemphasized, the growth in communication alone is not sustainable any longer [3]. On the other hand, recent advances in communication and information technologies have fueled a plethora of innovations in various areas, including *networking, caching and computing*, which have the potential to profoundly impact our society via the development of smart homes, smart transportation, smart cities[4], etc.

Therefore, the integration of networking, caching and computing into one system becomes a natural trend [5, 6]. By incorporating caching functionality into the network, the system can provide native support for highly scalable and efficient content retrieval, and meanwhile, duplicate content transmissions within the network can be reduced significantly. As a consequence, mobility, flexibility and security of the network can be considerably improved [7, 8]. On the other hand, the incorporation of computing functionality endows the network with powerful capabilities of data processing, hence enabling the execution of computationally intensive applications within the network. By offloading mobile devices' computation tasks (entirely or partially) to resource-rich computing infrastructures in the vicinity or in remote clouds, the task execution time can be considerably reduced, and the local resources of mobile devices, especially battery life, can be conserved, thus enriching the computing experience of mobile users [9]. Moreover, networking, caching and computing functionalities can complement and reinforce each other by interactions. For instance, some of the computation results can be cached for future use, thus alleviating the backhaul workload. On the other hand, some cached content can be transcoded into other versions to better suit specific user demands, thus economizing storage spaces and maximizing the utilization of caching.

Despite the potential vision of integrated networking, caching and computing systems, a number of significant research challenges remain to be addressed before widespread application of the integrated systems, including the latency requirement, bandwidth constraints, interfaces, mobility management, resource and architecture tradeoffs, convergence as well as non-technical issues such as governance regulations, etc. Particularly, the resource allocation issue, in which the management and allocation of three types of resources should be jointly considered to effectively and efficiently serve user requirements, is especially challenging. In addition, non-trivial security challenges are induced by a large number of intelligent devices/nodes with self adaptation/context awareness capabilities in integrated systems. These challenges need to be broadly tackled through comprehensive research efforts.

1.1 Overview

In this chapter, we present a brief overview on the history and current research progress of networking, computing and caching technologies in wireless systems.

1.1.1 Recent advances in wireless networking

The communication and networking technologies of the fifth generation (5G) wireless telecommunication networks are being standardized at full speed along the time line of International Mobile Telecommunications for 2020 (IMT-2020), which is proposed by the International Telecommunication Union (ITU) [10]. Consequently, unanimous agreement on the prospect of a new, backward-incompatible radio access technology is emerging in the industry.

Significant improvements on data rate and latency performance will be powered by *logical network slices* [11, 12] in the 5G communication environment. Furthermore, network flexibility will be greatly promoted by dynamic network slices, therefore providing the possibility of the emergence of a variety of new network services and applications. Due to the fact that the key problems of network slices are how to properly slice the network and at what granularity [11], it is not difficult to predict that a number of already existing technologies, such as software defined networking (SDN) [13] and network functions virtualization (NFV) [14], will be taken into consideration.

SDN can enable the separation between the data plane and the control plane, therefore realizing the independence of the data plane capacity from the control plane resource, which means high data rates can be achieved without incurring overhead upon the control plane [15, 16, 17]. Meanwhile, the separation of the data and control planes endows high programmability, low-complexity management, convenient configuration and easy troubleshooting to the network [18]. The controller in the control plane and the devices in the data plane are bridged by a well-defined application programming interface (API), of which a well known example is OpenFlow [19]. The instructions of a controller to the devices in the data plane are transmitted through flow tables, each of which defines a subset of the traffic and the corresponding action. Due to the advantages described above, SDN is considered promising on providing programmable network control in both wireless and wired networks [20, 21].

NFV presents the technology that decouples the services from the network infrastructures that provide them, therefore maximizing the utilization of network infrastructure by offering the possibility that services from different service providers can share the same network infrastructure [14, 22]. Furthermore, easy migration of new technology in conjunction with legacy technologies in the same network can be realized by isolating part of the network infrastructure [14].

On the other hand, heterogeneous network (HetNet) has been recently proposed as a promising network architecture to improve network coverage and link capacity [23, 24]. By employing different sizes of small cells and various radio access technologies in one macro cell, energy efficiency (EE) and spectral efficiency (SE) are significantly enhanced in scenarios such as shopping malls, stadiums and subway stations [24, 25, 26, 27]. But ultra-dense small cell deployment coexisting with legacy macro cells raise the problem of mutual interference, which calls for efficient radio resource allocation and management technologies [28, 29, 30].

Meanwhile, the Cloud Radio Access Network (C-RAN) is rising as a promising solution for reducing inter-cell interference, CAPital EXpenditure (CapEx) and OPerating EXpenditure (OpEx) [31]. Based on network centralization and virtualization, C-RAN is dedicated to promoting energy efficiency, coverage and mobility performances, while reducing the expense of network operation and deployment [31]. By pooling the baseband resources at the BaseBand Unit (BBU), which is located at a remote central office (not at the cell sites), C-RAN is able to provide the network operator with energy efficient operations, statistical multiplexing gains and resource savings [32, 33].

Massive multiple-input multiple-output (MIMO) technology was identified as a key 5G enabler by the European 5G project METIS in its final deliverable [34]. In a massive MIMO scenario, K pieces of user equipment (UE) are served on the same time-frequency resource by a base station (BS) equipped with M antennas, where $M \gg K$. The deployment of a large number of antennas at the BS leads to a particular propagation scenario named *favorable propagation*, where the wireless channel becomes near-deterministic since the BS-to-UE radio links become near-orthogonal to each other [35]. Recent advances in fields like 3D MIMO, hybrid beamforming (BF) and understanding of the asymptotic behavior of random matrices suggest that massive MIMO has the potential to bring unprecedented gains in terms of spectrum and energy efficiencies and robustness to hardware failures [34].

With the severe shortage in the currently utilized spectrum (sub-6 GHz), the utilization of higher frequency bands, e.g., millimeter-wave (mmWave), becomes a consensus of the communication community [36]. Working on 3–300 GHz bands, mmWave communications assisted by smart antenna arrays can track and transmit signals to high-speed targets over a long distance [37]. Since the physical antenna array size can be greatly reduced due to a decrease in wavelength, mmWave communications are appealing to large-scale antenna systems [38]. On the other hand, the detrimental effects of the high propagation loss in mmWave communications can be neatly compensated by the achievable high beamforming gains with massive MIMO [36, 39]. Therefore, mmWave communication assisted by massive MIMO is envisaged as a promising solution for future telecommunications.

Another way to deal with spectrum scarcity is to increase the utilization efficiency of the spectrum. Allowing two pieces of UE to perform direct data transmissions, device-to-device (D2D) communications enable a flexible reuse of radio resources, and therefore improve the spectral efficiency and ease core network data processing workloads [40, 41]. Moreover, the combination of D2D communications with information-centric networking (ICN) is worth considering. Despite the small-sized storage of UE, the ubiquitous caching capability residing in UE should not be neglected, due to the UE' pervasive distribution and ever-increasing storage sizes [42].

1.1.2 Caching

As described above, the spectral efficiency (SE) of wireless communication radio access networks has been increased greatly by ultra-dense small cell deployment. However, this has brought up another issue: the backhaul may become the bottleneck of the wireless communication system due to the tremendous and ever increasing amount of data being exchanged within the network [43]. On the other hand, building enough high speed backhauls linking the core network and the small cells which are growing in number could be exceptionally expensive. Being stimulated by this predicament, research efforts have been dedicated to caching solutions in wireless communication systems [44]. By storing Internet contents at infrastructures in radio access networks, such as base stations (BSs) and mobile edge computing (MEC) servers [45], caching solutions enable the reuse of cached contents and the alleviation of backhaul usage. Therefore, the problem has been transferred by caching solutions from intensive demands on backhaul connections to caching capability of the network.

However, the original intention of Internet protocols is providing direct connections between clients and servers, while the caching paradigm calls for a usage pattern based on distribution and retrieval of content. This contradiction has led to a degradation of scalability, availability and mobility [46]. To address this issue, Content Delivery Networks (CDNs) [48] and Peer-to-Peer (P2P) networks [49] have been proposed in application layers, as first attempts to confer content distribution and retrieval capabilities to networks by utilizing the current storage and processing network infrastructures.

CDNs employ clusters of servers among the Internet infrastructures and serve the UE with the replications of content that have been cached in those servers. The content requests generated by UE are transmitted through the Domain Name Servers (DNSs) of the CDN to the nearest CDN servers that hold the requested content, in order to minimize latency. The decision on which server is chosen to store the replication of content is made upon a constant monitoring and load balancing of data traffic in the network [50].

P2P networks rely on the storage and forwarding of replications of content by UE. Each UE can act as a caching server in P2P networks [50, 51]. In P2P networks, the sources of content are called *peers*. Instead of caching a complete content, each peer may store only a part of the content, which is called a *chunk*. Thus, the content request of a UE is resolved and directed to several peers by a directory server. Each peer will provide a part of the requested content upon receipt of the request.

Although CDN and P2P do give a solution for content distribution and retrieval, due to the fact that they solely operate on application layers and the commercial and technological boundaries that they are confined to, the performances of these two techniques are not ideal enough to fulfil the demands on network caching services [50].

Serving as an alternative to CDN and P2P networking, *Information-Centric Networking (ICN)* emphasizes information dissemination and content retrieval by operating a common protocol in a network layer, which can utilize current storage and processing network infrastructures to cache content. In general, depending on the location of caches, caching approaches of ICN can be categorized as *on-path caching* and *off-path caching* [52, 53]. On-path caching concerns the delivery paths when considering caching strategies, and hence are usually aggregated with the forwarding mechanisms of ICN. On the other hand, off-path caching solely focuses on the storage and delivery of content, regardless of the delivery path.

1.1.3 Computing

As the prevalence of smartphones is dramatically growing, new mobile applications, such as face recognition, natural language processing, augmented reality, etc. [1] are emerging. This leads to a constantly increasing demand on computational capability. However, due to size and battery life constraints, mobile devices tend to fail in fulfilling this demand. On the other hand, powered by network slicing, SDN and NFV, cloud computing (CC) [54] functionalities are incorporated into mobile communication systems, bringing up the concept of *mobile cloud computing (MCC)* [55], which leverages the rich computation resources in the cloud for user computation task execution, by enabling computation offloading through wireless cellular networks. After computation offloading, the cloud server will create a clone for each piece of user equipment (UE) individually, then the computation task of the UE will be performed by the clone on behalf of that UE. Along with the migration of computation tasks from UE to a resourceful cloud, we are witnessing a networking paradigm transformation from connection-oriented networking to content-oriented networking, which stresses data processing, storage and retrieval capability, rather than the previous criterion that solely focuses on connectivity.

Nevertheless, due to the fact that the cloud is usually distant from mobile devices, the low-latency requirements of some latency-sensitive (real-time) applications may not be fulfilled by cloud computing. Moreover, migration of a large amount of computation tasks over a long distance is sometimes infeasible and uneconomical. To tackle this issue, *fog computing* [56, 57, 47, 59] has been proposed to provide UE with proximity to resourceful computation servers. The terminology fog (From cOre to edGe) computing was first coined in 2012 by Cisco [60]. It is a distributed computing paradigm in which network entities with different computation and storage abilities and various hierarchical levels are placed within a short distance from the cellular wireless access network, connecting user devices to the cloud or Internet. It is worth noting that fog computing is not a replacement but a complement of cloud computing, due to the fact that the gist of fog computing is providing low-latency services to meet the demands of real-time applications, such as smart traffic monitoring, live streaming, etc. However, when the applications requiring a tremendous amount of computation or permanent storage are concerned, the fog computing infrastructures are only acting as gateways or routers for data redirection to the cloud computing framework [57].

The problems in wireless networks such as congestion of Internet connections, low-bandwidth and infeasibility of real-time applications can hardly be solved by simply and blindly boosting the underlying network bandwidth capacity. By analyzing real data collected from three hotels in Italy, the work in [59] shows that fog computing is competent in alleviating the effects of those problems. This analysis demonstrates that by deploying fog nodes and corresponding applications at the network edge, fog computing can proactively cache and manage up to 28.89 percent of the total traffic, which cannot be managed by conventional networking and caching approaches. Moreover, by imposing selective bandwidth limitations on specific connected devices, the fog nodes enable a dynamic and flexible management of the available bandwidth, bringing benefits to wireless network in terms of resource optimization and performance [59]. This real data based analysis can serve as a preliminary validation of the feasibility and significance of the incorporation of fog computing into wireless networks.

Also intended to solve the long-latency problem of mobile cloud computing, cloudlet-based mobile cloud computing is proposed [61]. Instead of utilizing the computation resource in a remote cloud, cloudlet-based mobile cloud computing can reduce data transmission delay by deploying computing servers/clusters within one-hop WiFi wireless access. However, there are two drawbacks associated with cloudlet based mobile cloud computing: First, cloudlet-based mobile cloud computing can hardly guarantee pervasive service due to the limited coverage of the WiFi access network which is usually employed in indoor environments. Second, due to the constraints on infrastructure size, the servers of cloudlet-based mobile cloud computing typically provide a small or medium amount of computation resources, which can hardly meet the computation requirement in the case of a large number of UE.

To serve as a complement of cloudlet-based mobile cloud computing, a new MCC paradigm similar to fog computing, named *mobile edge computing (MEC)*, is proposed [45, 62]. Being placed at the edge of radio access networks, MEC servers can provide sufficient computation resources in physical proximity to UE, which guarantees the fulfilment of the demand of fast interactive response by low-latency connections. Therefore, mobile edge computing is envisioned as a promising technique to offer agile and ubiquitous computation augmenting services for mobile devices, by conferring considerable computational capability to mobile communication networks [14].

1.2 Motivations and requirements

In this chapter, we first define the integrated system of networking, caching and computing in question, then we discuss the motivations of the integration, followed by the requirements of the integration of networking, caching and computing.

1.2.1 What is integration of networking, caching and computing?

In order to give a clear description of integrated networking, caching and computing, we first give the definitions of networking, caching and computing considered here. respectively.

■ Networking

The networking vector discussed in this chapter pertains to its capability to deliver data through networks with certain bandwidth and power. The measurement of this capability is usually data rate, whose unit is bits per second. The relationship among data rate, bandwidth and signal to noise power ratio is well demonstrated by Shannon's capacity formula.

■ Caching

The caching vector considered in wireless communication systems pertains to its ability to store a certain amount of data at the infrastructures in the network. The typical measurement of caching capability is the size of stored information, whose unit is bytes. Without altering the information flow in network, caching alleviates backhaul and enhances the capability of information delivery over a long term.

■ Computing

The computing vector under consideration in this chapter pertains to its capability to perform algebraic or logical operations over the information flows within the network. In contrary to the networking and caching vectors, the operation of the computing vector is intended to alter the information flow in the network. One of the typical measurements of computing capability is the number of information flows or nodes involved in the computing operation, which is called the degree of computing (DoC) [3].

Given the definitions of networking, caching and computing in wireless or mobile systems, we next move on to the discussion of the topic of this chapter, integrated networking, caching and computing. In this chapter, we consider integrated frameworks and technologies in wireless or mobile communication systems that take into consideration at least two aspects of *networking, caching and computing* as defined above, aiming at providing coherent and sustainable services to fulfil various mobile network user demands, such as high-data-rate communication, low-latency information retrieval and extensive computation resources. In other words, instead of solely focusing on the networking vector, we consider solutions concerning three network vectors, namely, networking, caching and computing, as shown in Figure 1.1. The acronyms IoT, AR and SMS in Figure 1.1 stand for Internet of Things, Augmented Reality and Short Message Service, respectively.

1.2.2   Why do we need integration of networking, caching and computing?

1.2.2.1   The growth of networking alone is not sustainable

Since the first application of commercial mobile cellular services around 1983, excessive research endeavors have been dedicated to the improvement of the communication capacity of mobile communication systems [3]. It is unanimously granted that in the early days of mobile communication systems, when the main functionality of the system was voice and texting services, improvement of throughput, latency and link capacity could significantly enhance the user experience, which in turn yielded greater profits for network operators. However, as years passed, the mobile communication systems have undergone

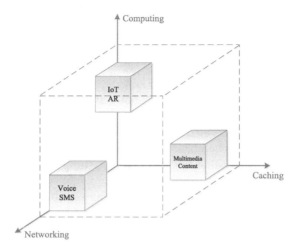

Figure 1.1: Schematic graph of integrated networking, caching and computing.

an evolution, in which their functionalities have grown. Especially in recent years, along with the growing prevalence of smartphones, new applications are emerging constantly.

Due to the fact that many multimedia applications rely on a tremendous amount of data, the demands of mobile network users on powerful storage services are rapidly increasing. According to the report of ARCchart [64], with the radically increasing camera resolution of mobile devices and the even faster growing size of data generated by such hardware advances, demands on mobile cloud storage are undergoing an exponential growth. To meet this user requirement, *mobile cloud storage* has risen and taken the place of the most widely used *cloud mobile media (CMM)* service, which is mainly provided by Google, Apple, Dropbox and Amazon today [1]. This service enables mobile network users to store video, music and other files in the cloud, to access files through any devices regardless of the source of the data, and to synchronize data in multiple devices. In order to employ a mobile cloud storage service on a large scale, it is imperative to ensure high integrity and availability of data, as well as user privacy and content security. It is clear that large scale deployment of a mobile cloud storage service will cause a significant increase in mobile data traffic, which in turn puts a demand on the integration of the storage functionality with ubiquitous and high-data-rate wireless networking functionality.

On the other hand, audio and video streaming based service, which typically needs excessive computation resources, is becoming one of the major services in mobile communication systems [1]. According to the technical report of CISCO [65], by the end of 2017, video traffic is expected to take over almost 80% of annual Internet traffic, whose total size would exceed 1.4 zettabytes *. However, the very limited computation resources provided by mobile devices can hardly meet the requirement raised by tasks such as encoding, transcoding and transrating in this category of service. Therefore, to solve this dilemma, the incorporation of powerful computing technologies like MCC and MEC into mobile communication systems becomes inevitable. By leveraging services of MCC and MEC, not only can the time consumption of computing be reduced, but also computing and backhaul costs can be reduced by caching popular videos through the caching functionality of MCC and MEC. Furthermore, the elasticity of computation resources provided by MCC/MEC servers is suitable for handling volatile peak demands in a cost-effective pattern.

In summary, the sole emphasis on networking capabilities such as connectivity and spectral efficiency can hardly fulfill user demands on the diversity of services, especially services that rely on intensive computation resources and enormous storage capabilities, and thus the growth of networking alone is unsustainable. The integration of networking, caching and computing is indispensable to achieve sustainability of mobile communication systems.

1.2.2.2 The benefits brought by the integration of networking, caching and computing

The incorporation of caching functionality into mobile communication systems enables information distribution and high-speed retrieval within the network. In-network caching and multicast mechanisms facilitate the transformation of data delivery pattern from direct server-client link to an information-centric paradigm, thus enhancing the efficiency and timeliness of information delivery to users [52]. Furthermore, network traffic reduction can also be realized by employing in-network caching in mobile communication systems. According to the trace-driven analysis in [66], with an in-network caching mechanism, three or more hops for 30% of the data packets in the network can be saved, or equivalently 2.02 hops on average can be saved. Moreover, up to 11% of users' requests for data can be retained within the requester domain.

On the other hand, the incorporation of computing functionality into mobile communication systems brings the benefits of access to the rich computation and storage resources in the cloud or the network edge, which enables the utilization of cutting edge multimedia techniques that are powered by much more intensive storage and computation capabilities than what the mobile devices can provide, thus offering richer and finer multimedia services than what primitive mobile applications can provide. Moreover, due to the fact that mobile applications enabled by MCC/MEC technologies can exploit not only the resources in the cloud or network edge, but also the unique resources in mobile devices, such as location information and camera data, these mobile

---

*1 zettabyte=10^{21} bytes.

applications are apparently more efficient and richer in experience than PC applications.

The integration framework makes it possible for users to access media and information through any platform, any device and any network. According to Juniper Research, the revenues of consumer cloud mobility services, whose initial prototypes are cloud based video and music caching and downloading services like Amazon's Cloud Drive and Apple's iCloud, reached 6.5 billion dollars per year by 2016 [67].

1.2.3 The requirements of integration of networking, caching and computing

In order to implement the integration of networking, caching and computing, several requirements need to be met.

1.2.3.1 Coexistence

In an integrated framework, it is important for the caching and computing infrastructures to coexist with the mobile communication infrastructures, to guarantee the seamless cooperation of the networking, caching and computing functionalities. Actually, coexistence is the prerequisite of the realization of integrated networking, caching and computing systems [10].

The infrastructure type, topology, QoS requirement, service type and security level are all different across these three functional vectors, and when they are integrated into one framework, they still need to bear all these different properties.

1.2.3.2 Flexibility

A certain degree of freedom in these three functional vectors is necessary. The flexibility depends on the degree of the integration of the three vectors [3]. A high level of integration may provide good multiplexing of the three vectors, seamlessness of cooperation between different functionalities and simplicity of implementation, while reducing the flexibility of resource customization in the integrated system. A low level of integration can lead to the reverse effects.

1.2.3.3 Manageability and programmability

The configuration, deployment and various resource allocation of the integrated framework bring up the requirements of manageability and programmability [3]. Only with a relatively high degree of manageability and programmability can the admission, maintenance and resource scheduling of the integrated system be viable. To realize programmability, the system needs to provide effective programming language, feasible interfaces and a secure programming environment with a considerable level of flexibility.

1.2.3.4 Heterogeneity

Different functional vectors, networking, caching and computing, typically operate in different network paradigms, and a heterogeneous network structure can facilitate the coexistence of different networking access manners [10].

1.2.3.5 Scalability

Since the number of mobile users and the complexity of mobile applications are increasing as time goes by, the integrated system of networking, caching and computing should have the capability to support a fluctuating number of UE and complexity of tasks.

1.2.3.6 Stability and convergence

The integration of networking, caching and computing should be stable enough to overcome errors, misconfiguration and other unstable situations. Also, any algorithm concerning the allocation of multiple resources should have a good convergence property.

1.2.3.7 Mobility

The integrated system should be able to support not only the geographical mobility of mobile users, but also the logical mobility of caching and computing functional vectors within the system.

1.2.3.8 Backward compatibility

Integrated systems should be able to exploit the existing communication network infrastructures up to the hilt. The caching and computing functionalities should be realized on the basis of utilizing existing networking resources. Therefore, from the economic point of view, the backward compatibility with legacy communication systems is greatly beneficial and vital. This means that the compatibilities on both interfaces and protocols should be taken into consideration when designing integrated systems.

1.3 Frameworks

In this chapter, we summarize the typical frameworks (architectures) of integrated systems of networking, caching and computing. The frameworks presented are based on existing studies, and reflect the basic ideas, mechanisms and components of integrated systems. Unfortunately, this brief summary may not cover all of the proposed architectures in the literature, due to the fact that each proposed architecture has its unique original intention, idea of design and working environment.

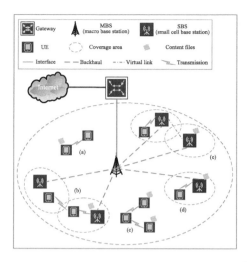

Figure 1.2: Networking-caching framework [8].

1.3.1 Caching-networking framework

In this subsection, we will stress on the study of [68], due to the fact that it provides a rough summary of various typical caching approaches within one framework. Without losing generality, we will also give a quick glance at frameworks proposed by other studies.

The framework proposed in [68] intends to provide a distributed caching and content delivery solution, taking into consideration diverse storage, content popularity distribution and user device mobility. In this framework, the communication and caching capabilities of small cell base stations (SBSs) and device-to-device (D2D) enabled mechanisms are jointly utilized to support a multilayer content caching and delivery framework. Both SBSs and user devices have the ability to store and transmit content.

The multilayer caching-networking framework is illustrated in Figure 1.2. Multiple small cells are deployed in one macro cell, and all the SBSs are connected to the macro cell base station (MBS) through backhaul links. The MBS is connected to a gateway of the core network via a high-speed interface. Each individual SBS and UE is equipped with a caching device to perform the caching function. According to current standards of the Third Generation Partnership Project (3GPP), X2 interfaces are solely intended for signaling between SBSs, and therefore cached content sharing between SBSs is not permitted. However, by utilizing inter-cell D2D communications, content cached in different small cells can be shared with each other.

As shown in Figure 1.2, this framework consists of the following five cached content delivery approaches.

1.3.1.1 D2D delivery (Fig. 1.2a)

Content requested by UE can be retrieved through D2D links from neighboring UE. A high density of UE can contribute to a high probability of successful retrieval of requested content. Moreover, the implementation of inter small cell D2D communications can further facilitate the utilization of content cached in adjacent small cells.

1.3.1.2 Multihop delivery via D2D relay (Fig. 1.2b)

D2D communication links can serve as a relay for content delivery from other sources to the destination UE. This approach enables a broader range of caching and delivery.

1.3.1.3 Cooperative D2D delivery (Fig. 1.2c)

If multiple UE held the content requested by another UE, they can cooperatively transmit the content to the destination UE through a D2D multiple-input multiple-output (MIMO) technology to accelerate the transmission process. Moreover, if video coding techniques such as multiple description coding and scalable video coding are taken into consideration during video data deliveries, more benefits may be gained.

1.3.1.4 Direct SBS delivery (Fig. 1.2d)

If the requested content was cached in the associated SBS, the SBS can transmit it directly to the destination UE. Despite the interference incurred during the retrieval and transmission processes, the latency can be lower than multihop transmission approaches.

1.3.1.5 Cooperative SBS delivery (Fig. 1.2e)

If the content requested by a certain UE was stored neither in its nearby UE nor in its associated SBS, it may retrieve and demand the content from an SBS in a neighboring small cell. In this case, the associated SBS will require the demanded content from neighboring SBSs through virtual links, which means these two SBSs are able to communicate with each other via their respective backhauls to the MBS.

Three function modules, namely, REQProxy, ContentServer, and Gate-Server, are designed in the application layer to enable the above five content delivery approaches. REQProxy recognizes content requests from UE and transmits the requests to ContentServer. ContentServer performs content retrieving and forwarding. GateServers are embedded in SBSs, and they send content requests that cannot be handled locally to other SBSs or remote servers.

The signaling interaction procedures of this framework are shown in

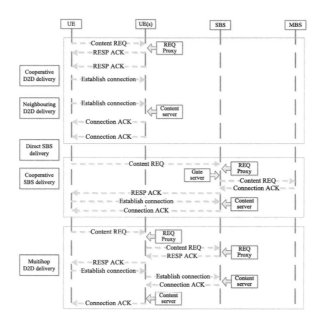

Figure 1.3: Signaling interaction of the networking-caching framework [68].

Figure 1.3. In the following, three delivery approaches are taken as examples to demonstrate the signaling interaction procedures.

Cooperative D2D delivery (Fig. 1.2c). When any content is needed by a certain UE, the Content REQ (content request) is sent by the UE to nearby UE. Then each individual receiver's REQProxy will contact its own ContentServer to check whether the requested content is cached by itself. If so, the REQProxy of the receiver will respond to the REQProxy of the requester with a RESP ACK (acknowledgment). After that, the content transmission link will be established with another two packets, Establish Connection and Connection ACK. This signaling interaction procedure is also applicable in the process of plain D2D delivery (Fig. 1.2a) and Direst SBS delivery (Fig. 1.2d), where the associated SBS responds to the requester with RESP ACK.

Cooperative SBS delivery (Fig. 1.2e). In this approach, the requester UE sends a Content REQ to its associated SBS, then the SBS retransmits this Content REQ to the MBS, through GateServer. The virtual link between the SBS holding the requested content and the requester's associated SBS will be established. Then the content will be transmitted to and cached at the associated SBS and eventually delivered to the requester UE.

Multihop delivery via D2D relay (Fig. 1.2b). In this approach, the relay UE needs to establish a link with its associated SBS. When the requested content is ready to be transmitted, the SBS will send a connection ACK to the relay UE in order to establish transmission connection.

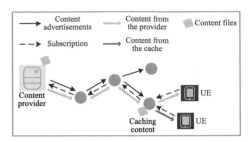

Figure 1.4: The information-centric networking [8].

An important feature of 5G mobile networking could be recognizing the content as the first-class citizen in the network [8]. To fulfil this requirement, the conventional content delivery networking (CDN) [69, 70] has distributed further, incorporating content-aware and information-centric caching technologies, and forming the so-called information-centric networking (ICN) [71], which has the significant advantages of natively facilitating efficient and scalable content retrieval and enhancing the mobility and security of the network. As an important native feature of ICN, in-network caching has the capability of significantly reducing duplicate data transmissions in the network. ICN-based air caching technology is considered a promising candidate to support the implementation of the SDN-based 5G mobile network [71].

The study of [8] proposes an information-centric wireless network (ICN) virtualization framework, which involves content-level slicing functionality. As shown in Figure 1.4, in contrast with Internet Protocol (IP), which adopts a host-based conversation model (i.e., establish and maintain connection between hosts before actual data delivery) and follows a source-driven data delivery approach (i.e., path is initiated by sender to receiver), ICN follows a receiver-driven communication principle (i.e., path is initiated by receiver to sender), and the data transmission follows the reverse way. As can be seen in Figure 1.4, network nodes in ICN can cache the content that flows through it and disseminate the content to other UE. Instead of the reachability of the hosts and the maintenance of the connectivity between hosts, the main concern of ICN is retrieval and dissemination of information.

The authors of [8] present a content-level slicing in their proposed virtualized ICN framework, which can be considered an extension of dynamic content access and sharing. The physical content (cache) is sliced through space multiplexing, time multiplexing and so forth, then allocated to different service providers [72]. In this approach, several services share a number of physical contents cached in the network, and each content is sliced into a number of virtual contents. Each service can utilize one or several of the virtual contents without the knowledge of other virtual contents. The number of virtual copies of the physical contents can depend on the popularity of the content, i.e., more popular physical content is sliced into more virtual contents.

Conceptually speaking, content-level slicing is a realization of caching sharing and dynamic access in a virtualized networking environment.

Not only the reduction of network data traffic but also the improvement of quality of experience can be realized by moving the frequently requested content toward the network edge closer to UE [71]. The study of [73] proposes a framework of caching popular video clips in the RAN. The concept of FemtoCaching is proposed in [74], and the video content caching and delivery approach with the facilitation of distributed caching helpers in femtocell networks is discussed. The research conducted in [71] studies current caching technologies and discusses caching in 5G networking, including radio access network caching and evolved packet core network caching. Then an edge caching approach (on the edge of RAN) based on information-centric networking is designed. The paper discusses architecture design issues on caching within evolved packet core (EPC), caching at RAN and content-centric networking (CCN) based caching.

As the popularity of social media is increasing dramatically, online social networks (OSN) are starting a new revolution in how people communicate and share information. In centralized OSN, the network has a central repository of all user data and imposes control regarding how user data will be accessed. This form of architecture has the advantage of efficient dissemination of social updates. However, users do not have control over their personal data, and are therefore constantly exposed to potential privacy violations. To address this issue, the architecture of distributed online social networks (DOSN) has been proposed [75, 76]. DOSN does not have a central repository and hence guarantees the user privacy, but faces the problem of inefficient dissemination of social updates. The work in [75] introduces Social Caches as a way to alleviate the peer-to-peer traffic in DOSNs, and the social caches are supposed to act as local bridges among friends to provide efficient information delivery. Furthermore, a social communication model called Social Butterfly has been proposed to utilize social caches. However, the selection of social caches demands full knowledge of the whole social graph. Therefore the authors in [76] propose a number of distributed social cache selection algorithms.

Due to the fact that stable D2D links can be easily maintained among people who have close social relationships, social-aware device-to-device (D2D) communication is proposed to complement the online social network communication. The work in [77] proposes a fog radio access network (FRAN) supported D2D communication architecture, in which the UE can download files either from a centralized baseband unit (BBU) pool through distributed remote radio heads (RRHs), or from other UE through D2D links. In [78], a context-aware framework for optimizing resource allocation in small cell networks with D2D communication is proposed. Both the wireless and social context of wireless users are utilized to optimize resource allocation and improve traffic offload.

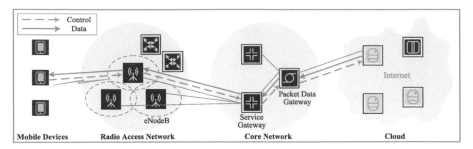

Figure 1.5: Cloud mobile media framework, demonstrating data and control flows [1].

1.3.2 Computing-networking framework

In this subsection, we will consider the frameworks of Cloud Mobile Media (mobile cloud computing) and mobile edge computing as two representatives to give a general profile of the architecture of the computing-networking paradigm.

1.3.2.1 Cloud mobile media

The overall architecture of cloud mobile media (CMM) is depicted in Figure 1.5, including end to end control and data flows between Internet cloud servers and mobile devices [1]. Enabled by cloud mobile media, applications on mobile devices are able to utilize intensive and elastic computing and caching resources in remote clouds (typically located in the Internet), including private, public and federated (hybrid) cloud types. The user interfaces of CMM applications used for command input are provided by user devices, including touchscreens, gestures, voices and texts. The subsequent control commands are then transmitted uplink via cellular radio access networks (RAN) to gateways located in core networks (CN), and eventually to the serving clouds. After that, the clouds perform data processing or content retrieval using computing or caching resources in cloud servers, in accordance with user control commands. Then the results are sent back to user devices downlink through CN and RAN.

Please note that most of the applications follow the data and control flows shown in Figure 1.5, but there exist some exceptions that slightly deviate from them. For instance, the control flows in cloud-based media analytics are not always initiated by mobiles devices, and this type of application may collect data from both the cloud and the mobile devices in order to perform analytics for other types of CMM applications.

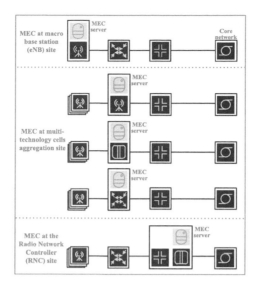

Figure 1.6: MEC server deployment scenarios [45].

1.3.2.2 *Mobile edge computing*

As described in previous sections of this chapter, MEC serves as a complement to mobile cloud computing, and the intention of MEC is reducing the high data transmission latency of MCC by providing proximity to computation resources at the edge of a radio access network. The basic mechanism of MEC is quite similar to that of MCC. They both allow UE to offload computation tasks to servers that execute the tasks on behalf of the UE, then the results are returned to the UE.

MEC servers can be deployed at various locations within the edge of radio access networks. According to the report of [45], the MEC server can be deployed at the macro base station (eNB) site, at the multi-technology cell aggregation site, or at the radio network controller (RNC) site. The deployment scenarios are depicted in Figure 1.6. The multi-technology cell aggregation site can be located either at an indoor environment in an enterprise, such as a university center or airport, or at an outdoor environment to control several local multi-technology access points providing coverage over certain public premises. Direct delivery of fast, locally relevant services from base station clusters to UE can be enabled by this option of deployment.

1.3.3 *Caching-computing framework*

In recent years big data has attracted more and more attention due to the increasing diversity of mobile applications and the tremendous amount of data streaming in mobile networks. In this context, data management/processing and data caching have emerged as two key issues in this field. The study

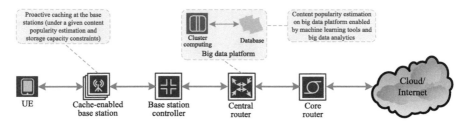

Figure 1.7: Caching-computing architecture [79].

of [79] proposes a caching-computing framework in which cached content is moved from the cloud to the RAN edge, and a big data platform powered by machine learning and big data analytics is deployed at the core site for content selection.

As illustrated in Figure 1.7, the caching platform at small cell base stations (SBSs) and the computation and execution of the content prediction algorithms at the core site are two key components of the proposed architecture. A big data platform is employed at the core site to perform tracking and prediction of users' demands and behavior for content selection decisions, and SBSs in the RAN are equipped with caching units to store the strategic content selected by the big data platform. The reason for employing the big data platform is that the content cached in SBSs can hardly cover all the users' content demands, due to the limited storage capacity of the SBSs. In order to yield optimal performance, it is important to select the most popular content (most frequently requested content based on estimation) and to decide the content cache placement at specific small cells based on backhauls, rate requirements, content sizes and so forth.

The big data platform portrayed in Figure 1.7 is used for sorting users' data traffic and extracting statistical information for proactive caching decisions. To achieve this goal, the following requirements should be met by the big data platform.

Enormous amount of data processing in a short time period. Making proactive caching decisions requires the capability of processing a large amount of data and drawing intelligent insights in a short time.

Cleansing, parsing and formatting data. In order to perform statistical analysis and machine learning, the data should be cleaned in advance. The malfunctioning, inappropriate and inconsistent packets involved in the raw data should be eliminated before any type of data processing. Then the relevant fields should be extracted from the raw data. Finally, the parsed data should be encoded accordingly.

Data analysis. A variety of data analysis technologies can be applied in the analysis of the header and/or payload information of both control and data planes. The purpose of data analysis is finding the relationship between

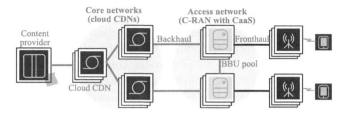

Figure 1.8: Architecture of cloud content-centric mobile networking [80].

control/data packets and the requested content and then predicting spatial-temporal user behavior.

Statistical analysis and visualizations. The outcomes of the data analysis can be stored and reused for further statistical analysis. Moreover, the outcomes can be formatted as graphs or tables for enhancement of readability and ease of comprehension.

In [80], the authors propose a cloud content-centric mobile networking (CCMN) architecture, which incorporates cloud computing into both core network caching and radio access network caching. In this architecture, the conventional content delivery network (CDN) migrates into the cloud, forming a cloud CDN which leverages the merits of both cloud computing and CDN. As an application of cloud computing, the cloud CDN architecture inherits most of the advantages of cloud computing, and moreover, it can offer sufficient elasticity for handling flash crowd traffic. However, the content proximity to UE and the coverage of cloud CDN are not as ideal as those of conventional CDNs.

The merging of cloud computing and RAN forms the so-called cloud RAN (C-RAN), whose most significant innovation is that the computationally intensive baseband processing tasks are executed by a centralized cloud baseband unit (BBU) pool, instead of the traditional distributed baseband processing devices at the base station sites [81]. By adding the content caching functionality into conventional C-RAN, C-RAN with caching as a service (CaaS) is considered a promising approach to overcome the limitations of core network caching, due to the fact that the content is moved from a core network to a RAN, and hence is much closer to UE. In C-RAN with CaaS, the BBU pool can offer larger caching space than that of BSs, and the centralized BBU pool can facilitate solving the distributed content placement problem over geographically scattered BSs. Nevertheless, C-RAN with CaaS doesn't do well in fronthaul traffic mitigation.

The overall architecture of the cloud content-centric mobile networking (CCMN) is depicted in Figure 1.8, which shows the assistant role of cloud computing in both core network caching and RAN caching.

A similar framework called fog RAN (F-RAN), combining edge caching and C-RAN, has been recently advocated [82, 83, 84]. In F-RAN, edge nodes are equipped with caching capabilities, and simultaneously are controllable by a central cloud processor as in C-RANs. C-RANs can guarantee high spectral efficiency due to cooperative cloud-based transmission, while the latency is not ideal because of fronthaul transmission. On the other hand, edge caching can provide low-latency deliveries of popular content, but suffers from poor interference mitigation performance due to decentralized baseband processing at the edge nodes. By leveraging features of both edge caching and C-RAN, F-RAN can benefit from both low-latency deliveries of content and centralized baseband processing.

1.3.4   Caching-computing-networking framework

In this subsection, we generally consider two categories of networking-caching-computing frameworks, namely, networking-caching-computing convergence and networking and computing assisted caching. The former refers to the framework in which networking, caching and computing functionalities coalesce with each other to form a comprehensive structure, providing UE with versatile services. The latter means in this type of frameworks networking and computing capabilities of the network elements are functioning to provide supports to caching functionality, improving the quality of caching services of the networks.

1.3.4.1   Networking-caching-computing convergence

The synergy of networking, caching and computing is not a trivial step. First, different resource accesses require unique communication protocols, and therefore the convergence of these three types of resource accesses poses a requirement on new protocol design. Second, interactions between different types of resources need appropriate interfaces on network devices, which means redesign of network devices is necessary.

The authors in [5] propose a framework incorporating networking, caching and computing functionalities under the control and management of an SDN control plane, aiming at providing energy-efficient information retrieval and computing services in green wireless networks. A software-defined approach is employed in the integrated framework, in order to leverage the separation of the control plane and the data plane in SDN, which help guarantee flexibility of the solution. Three types of radio access networks are considered in this chapter, namely, cellular networks, wireless local area networks (WLANs) and worldwide interoperability for microwave access (WiMAX). Each network node is equipped with caching and computing capabilities, and is under the control of a controller, which is in charge of the topology of the whole wireless network and the packet forwarding strategies. Thus, this framework can provide UE with caching, computing, as well as communication services.

1.3.4.2 *Networking and computing assisted caching*

Caching systems face three primary issues: *where to cache, what and how to cache,* and *how to retrieve.* To answer the question *where to cache,* identification of eligible candidates for caching the content is needed. In other words, it is necessary to clarify the criteria of a node being an information hub in the network. Following that, the question *what to cache* with respect to the popularity and availability of the content is posed. It is also imperative to decide which nodes should keep which content for redundant caching avoidance. After the content is appropriately cached in the network, the final question *how to retrieve* requires an answer.

In order to accommodate the dramatically increasing content requests generated by a huge number of UE in geographical proximity in vehicular networks, the study in [85] advocates the adoption of a networking and computing assisted caching paradigm in information-centric networking architecture, for ICN can provide the decoupling of content providers and consumers, and innetwork caching functionality is intrinsically supported by ICN. Therefore, the paper proposes a socially aware vehicular information-centric networking model, in which mobile nodes such as vehicles are equipped with communication, caching and computing capabilities, and they are responsible for computing their eligibility for caching and delivering content to other nodes (i.e., the relative importance of vehicles in the network, with respect to user interests, spatial temporal availability and neighbourhood connectivity). Furthermore, each node also bears the mission of computing its eligibility for being an information hub in the network. This has addressed the *where to cache* question. For the *what and how to cache* question, the decision is made on the computation results regarding content popularity, availability and timeliness, taking into consideration cooperative caching schemes. As for the *how to retrieve* question, the paper presents a social content distribution protocol for vehicles to relay and retrieve cached content, in order to enhance information reachability.

1.3.5 A use case

Here we provide a use case to further elaborate the interactions of the three dimensions in the integrated framework.

In this use case, we consider wireless network virtualization (WNV) technology, which realizes physical resource sharing between different virtual networks by creating virtual BSs upon physical infrastructures. A virtual BS could be a BS, access point (AP) or road side unit (RSU) in the physical wireless network. As shown in Figure 1.9, a cluster of virtual BSs is connected to one MEC server, and they are all connected to the Internet through the core network. A cache (storage facility) is deployed at each virtual BS.

Joe finds an interesting monument when he is visiting Ottawa. With his augmented reality (AR) glasses, Joe notices that there is an introductory video

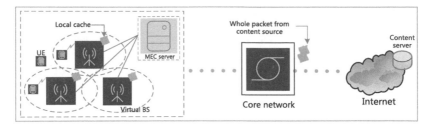

Figure 1.9: A use case of the integrated system of networking, caching and computing.

about this monument. Being interested in this video, he generates a video content request via his AR glasses to a virtual BS. According to the description of the video content and the information about the user, the virtual BS will check whether its associated cache has the requested content. If yes, the virtual BS will further examine if the version of its cached video content matches the user's device. If still yes, the virtual BS will directly send the requested content to the user from the cache. If no, the virtual BS will extract the current video content and initiate a computation task according to the size of the content including the input data, codes and parameters, as well as the number of CPU cycles that is needed to accomplish the computing/transcoding task. Then the virtual BS will transmit the task to the MEC server to execute the computation. After the computation is finished, the virtual BS sends the transcoded video content to the user. If the virtual BS cannot find any version of the requested video content in the cache, the virtual BS has to retrieve the content from the Internet, and this will inevitably consume some of the backhaul resources. Upon the arrival of the new video content at the virtual BS, the virtual BS can choose whether to store the content in its cache or not. The procedure is shown in Figure 1.10.

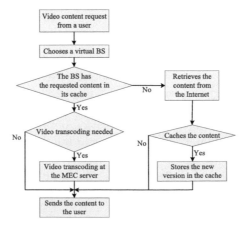

Figure 1.10: The processing procedure of the use case.

The above procedure is for downlink streaming video. Similarly, the procedure for uplink streaming video can be considered as follows. When Joe finds an interesting monument, he takes a video of it, puts some comments on it, and streams it to a remote server via a virtual BS. The virtual BS extracts the video content, and estimates the possibility that other people want to see this video. If the possibility is high, the virtual BS will cache this video locally, so that other people who are interested in this video can obtain this video from its cache directly, instead of requesting it from the remote server. As thus, backhaul resources can be saved and propagation latency can be reduced. If the cached version does not match the new user's device, transcoding will be performed by the MEC server.

Although the above use case focuses on tourism services, similar use cases with similar requirements can be derived for other services and applications, such as transportation, education, healthcare, etc.

References

[1] S. Wang and S. Dey, "Adaptive mobile cloud computing to enable rich mobile multimedia applications," *IEEE Trans. Multimedia*, vol. 15, no. 4, pp. 870–883, June 2013.

[2] S. Abolfazli, Z. Sanaei, E. Ahmed, A. Gani, and R. Buyya, "Cloudbased augmentation for mobile devices: motivation, taxonomies, and open challenges," *IEEE Communications Surveys & Tutorials*, vol. 16, no. 1, pp. 337–368, 2014.

[3] H. Liu, Z. Chen, and L. Qian, "The three primary colors of mobile systems," *IEEE Commun. Mag.*, vol. 54, no. 9, pp. 15–21, Sept. 2016.

[4] Y. He, F. R. Yu, N. Zhao, V. C. M. Leung, and H. Yin, "Softwaredefined networks with mobile edge computing and caching for smart cities: A big data deep reinforcement learning approach," *IEEE Commun. Mag.*, vol. 55, no. 12, Dec. 2017.

[5] R. Huo, F. Yu, T. Huang, R. Xie, J. Liu, V. Leung, and Y. Liu, "Software defined networking, caching, and computing for green wireless networks," *IEEE Commun. Mag.*, vol. 54, no. 11, pp. 185–193, Nov. 2016.

[6] Q. Chen, F. R. Yu, T. Huang, R. Xie, J. Liu, and Y. Liu, "An integrated framework for software defined networking, caching and computing," *IEEE Network*, vol. 31, no. 3, pp. 46–55, May 2017.

[7] C. Fang, F. R. Yu, T. Huang, J. Liu, and Y. Liu, "A survey of green information-centric networking: Research issues and challenges," *IEEE Comm. Surveys Tutorials*, vol. 17, no. 3, pp. 1455–1472, 2015.

[8] C. Liang, F. Yu, and X. Zhang, "Information-centric network function virtualization over 5G mobile wireless networks," *IEEE Netw.*, vol. 29, no. 3, pp. 68–74, May 2015.

[9] Z. Sanaei, S. Abolfazli, A. Gani, and R. Buyya, "Heterogeneity in mobile cloud computing: Taxonomy and open challenges," *IEEE Commun. Surv. Tuts.*, vol. 16, no. 1, pp. 369–392, 2014.

[10] S. Andreev, O. Galinina, A. Pyattaev, J. Hosek, P. Masek, H. Yanikomeroglu, and Y. Koucheryavy, "Exploring synergy between communications, caching, and computing in 5G-grade deployments," *IEEE Commun. Mag.*, vol. 54, no. 8, pp. 60–69, Aug. 2016.

[11] M. Richart, J. Baliosian, J. Serrat, and J. Gorricho, "Resource slicing in virtual wireless networks: A survey," *IEEE Trans. Netw. and Service Management*, vol. 13, no. 3, pp. 462–476, Aug. 2016.

[12] X. Zhou, R. Li, and T. Chen, "Network slicing as a service: enabling enterprises own software-defined cellular networks," *IEEE Commun. Mag.*, vol. 54, no. 7, pp. 146–153, Jul. 2016.

[13] N. Bizanis and F. Kuipers, "SDN and virtualization solutions for the internet of things: A survey," *IEEE Access*, vol. 4, pp. 5591–5606, Sept. 2016.

[14] C. Liang and F. Yu, "Wireless network virtualization: A survey, some research issues and challenges," *IEEE Commun. Surv. Tuts.*, vol. 17, no. 1, pp. 358–380, Aug. 2014.

[15] C. Lai, R. Hwang, H. Chao, M. Hassan, and A. Alamri, "A buffer-aware http live streaming approach for sdn-enabled 5g wireless networks," *IEEE Netw.*, vol. 29, no. 1, pp. 49–55, Jan. 2015.

[16] P. Agyapong, M. Iwamura, D. Staehle, W. Kiess, and A. Benjebbour, "Design considerations for a 5g network architecture," *IEEE Commun. Mag.*, vol. 52, no. 11, pp. 65–75, Nov. 2014.

[17] Q. Yan, F. R. Yu, Q. Gong, and J. Li, "Software-defined networking (SDN) and distributed denial of service (DDoS) attacks in cloud computing environments: A survey, some research issues, and challenges," *IEEE Commun. Survey and Tutorials*, vol. 18, no. 1, pp. 602–622, 2016.

[18] D. Kreutz, F. Ramos, P. Verissimo, C. Rothenberg, S. Azodolmolky, and S. Uhlig, "Software-defined networking: A comprehensive survey," *Proceedings of the IEEE*, vol. 103, no. 1, pp. 14–76, Jan. 2015.

[19] N. McKeown, T. Anderson, H. Balakrishnan, G. Parulkar, L. Peterson, J. Rexford, S. Shenker, and J. Turner, "Openflow: Enabling innovation in campus networks," *SIGCOMM Comput. Commun. Rev.*, vol. 38, no. 2, pp. 69–74, Mar. 2008.

[20] J. Liu, S. Zhang, N. Kato, H. Ujikawa, and K. Suzuki, "Device-to-device communications for enhancing quality of experience in software defined multi-tier lte-a networks," *IEEE Network*, vol. 29, no. 4, pp. 46–52, Jul. 2015.

[21] L. Cui, F. R. Yu, and Q. Yan, "When big data meets software-defined networking: SDN for big data and big data for SDN," *IEEE Network*, vol. 30, no. 1, pp. 58–65, 2016.

[22] C. Liang and F. Yu, "Virtual resource allocation in information-centric wireless virtual networks," in *Proc. IEEE International Conference on Communications (ICC)*, Jun. 2015, pp. 3915–3920.

[23] J. Hoadley and P. Maveddat, "Enabling small cell deployment with Het-Net," *IEEE Wireless Commun. Mag.*, vol. 19, no. 2, pp. 4–5, Apr. 2012.

[24] R. Xie, F. R. Yu, H. Ji, and Y. Li, "Energy-efficient resource allocation for heterogeneous cognitive radio networks with femtocells," *IEEE Trans. Wireless Commun.*, vol. 11, no. 11, pp. 3910–3920, Nov. 2012.

[25] M. Agiwal, A. Roy, and N. Saxena, "Next generation 5G wireless networks: A comprehensive survey," *IEEE Commun. Surv. Tuts.*, vol. 18, no. 3, pp. 1617–1655, 2016.

[26] T. Olwal, K. Djouani, and A. Kurien, "A survey of resource management toward 5G radio access networks," *IEEE Commun. Surv. Tuts.*, vol. 18, no. 3, pp. 1656–1686, 2016.

[27] L. Ma, F. Yu, V. C. M. Leung, and T. Randhawa, "A new method to support UMTS/WLAN vertical handover using SCTP," *IEEE Wireless Commun.*, vol. 11, no. 4, pp. 44–51, Aug. 2004.

[28] F. Yu and V. Krishnamurthy, "Optimal joint session admission control in integrated WLAN and CDMA cellular networks with vertical handoff," *IEEE Trans. Mobile Computing*, vol. 6, no. 1, pp. 126–139, Jan. 2007.

[29] S. Bu and F. R. Yu, "Green cognitive mobile networks with small cells for multimedia communications in the smart grid environment," *IEEE Trans. Veh. Tech.*, vol. 63, no. 5, pp. 2115–2126, Jun. 2014.

[30] W. Cheng, X. Zhang, and H. Zhang, "Statistical-QoS driven energy efficiency optimization over green 5g mobile wireless networks," *IEEE J. Sel. Areas Commun.*, vol. PP, no. 99, pp. 1–16, Mar. 2016, IEEE Early Access Articles.

[31] G. Huang and J. Li, "Interference mitigation for femtocell networks via adaptive frequency reuse," *IEEE Trans. Veh. Tech.*, vol. 65, no. 4, pp. 2413–2423, Apr. 2016.

[32] A. Checko, "Cloud RAN for mobile networks–a technology overview," *IEEE Commun. Surv. Tuts.*, vol. 17, no. 1, pp. 405–426, 2015.

[33] Y. Cai, F. R. Yu, and S. Bu, "Cloud computing meets mobile wireless communications in next generation cellular networks," *IEEE Network*, vol. 28, no. 6, pp. 54–59, Nov. 2014.

[34] Z. Yin, F. R. Yu, S. Bu, and Z. Han, "Joint cloud and wireless networks operations in mobile cloud computing environments with telecom operator cloud," *IEEE Trans. Wireless Commun.*, vol. 14, no. 7, pp. 4020–4033, July 2015.

[35] G. Fodor, N. Rajatheva, W. Zirwas et al., "An overview of massive mimo technology components in metis," *IEEE Commun. Mag.*, vol. 55, no. 6, pp. 155–161, Jul. 2017.

[36] K. Prasad, E. Hossain, and V. Bhargava, "Energy efficiency in massive mimo-based 5g networks: Opportunities and challenges," *IEEE Wireless Commun.*, vol. 24, no. 3, pp. 86–94, Jan. 2017.

[37] L. You, X. Gao, G. Li, X. Xia, and N. Ma, "BDMA for millimeter-wave/terahertz massive mimo transmission with per-beam synchronization," *IEEE J. Sel. Areas Commun.*, vol. 35, no. 7, pp. 1550–1563, Jul. 2017.

[38] L. Kong, M. Khan, F. Wu, G. Chen, and P. Zeng, "Millimeter-wave wireless communications for iot-cloud supported autonomous vehicles: Overview, design, and challenges," *IEEE Commun. Mag.*, vol. 55, no. 1, pp. 62–68, Jan. 2017.

[39] S. Han, C. I., Z. Xu, and C. Rowell, "Large-scale antenna systems with hybrid analog and digital beamforming for millimeter wave 5g," *IEEE Commun. Mag.*, vol. 53, no. 1, pp. 186–194, Jan. 2015.

[40] N. Ishikawa, R. Rajashekar, S. Sugiura, and L. Hanzo, "Generalized spatial-modulation-based reduced-rf-chain millimeter-wave communications," *IEEE Trans. Veh. Technol.*, vol. 66, no. 1, pp. 879–883, Jan. 2017.

[41] R. Ma, N. Xia, H. Chen, C. Chiu, and C. Yang, "Mode selection, radio resource allocation, and power coordination in D2D communications," *IEEE Wireless Commun.*, vol. 24, no. 3, pp. 112–121, Feb. 2017.

[42] Y. Cai, F. R. Yu, C. Liang, B. Sun, and Q. Yan, "Software defined device-to-device (D2D) communications in virtual wireless networks with imperfect network state information (NSI)," *IEEE Trans. Veh. Tech.*, no. 9, pp. 7349–7360, Sept. 2016.

[43] K. Wang, F. Yu, H. Li, and Z. Li, "Information-centric wireless networks with virtualization and D2D communications," *IEEE Wireless Commun.*, vol. 24, no. 3, pp. 104–111, Jan. 2017.

[44] Y. Zhou and W. Yu, "Optimized backhaul compression for uplink cloud radio access network," *IEEE J. Sel. Areas Commun.*, vol. 32, no. 6, pp. 1295–1307, Jun. 2014.

[45] A. Ioannou and S. Weber, "A survey of caching policies and forwarding mechanisms in information-centric networking," *IEEE Commun. Surv. Tuts.*, vol. PP, no. 99, pp. 1–41, 2016, IEEE Early Access Articles.

[46] ETSI, "Mobile-edge computing: Introductory technical white paper," *ETSI White Paper*, Sept. 2014.

[47] V. Jacobson, D. Smetters, J. Thornton, M. Plass, N. Briggs, and R. Braynard, "Networking named content," in *Proc. 5th International Conference on Emerging Networking Experiments and Technologies (CoNEXT09)*, Rome, Italy, Dec. 2009, pp. 1–12.

[48] R. Buyya and M. Pathan, *Content Delivery Network*. Berlin: Springer, 2008.

[49] S. Androutsellis and D. Spinellis, "A survey of peer-to-peer content distribution technologies," *ACM Computing Surveys (CSUR)*, vol. 36, no. 4, pp. 335–371, Dec. 2004.

[50] A. Passarella, "A survey on content-centric technologies for the current internet: CDN and p2p solutions," *Elsevier Journal on Computer Communications*, vol. 35, no. 1, pp. 1–32, Jan. 2011.

[51] E. Lua, J. Crowcroft, M. Pias, R. Sharma, and S. Lim, "A survey and comparison of peer-to-peer overlay network schemes," *IEEE Commun. Surv. Tuts.*, vol. 7, no. 2, pp. 72–93, Apr. 2005.

[52] B. Ahlgren, C. Dannewitz, C. Imbrenda, D. Kutscher, and B. Ohlman, "A survey of information-centric networking," *IEEE Commun. Mag.*, vol. 50, no. 7, pp. 26–36, July 2012.

[53] G. Xylomenos, C. Ververidis, V. Siris, N. Fotiou, C. Tsilopoulos, X. Vasilakos, K. Katsaros, and G. Polyzos, "A survey of informationcentric networking research," *IEEE Commun. Surv. Tuts.*, vol. 16, no. 2, pp. 1–26, July 2013.

[54] K. Djemame, D. Armstrong, J. Guitart, and M. Macias, "A risk assessment framework for cloud computing," *IEEE Trans. Cloud Computing*, vol. 4, no. 3, pp. 265–278, Jul. 2016.

[55] A. Khan, M. Othman, S. Madani, and S. Khan, "A survey of mobile cloud computing application models," *IEEE Commun. Surv. Tuts.*, vol. 16, no. 1, pp. 393–413, 2014.

[56] M. Zhanikeev, "A cloud visitation platform to facilitate cloud federation and fog computing," *Computer*, vol. 48, no. 5, pp. 80–83, May 2015.

[57] S. Sarkar, S. Chatterjee, and S. Misra, "Assessment of the suitability of fog computing in the context of Internet of things," *IEEE Trans. Cloud Computing*, vol. PP, no. 99, 2015, online.

[58] R. Gargees, B. Morago, R. Pelapur, D. Chemodanov, P. Calyam, Z. Oraibi, Y. Duan, G. Seetharaman, and K. Palaniappan, "Incident supporting visual cloud computing utilizing software-defined networking," *IEEE Trans. on Circuits and Systems for Video Technology*, pp. 1–15, 2016, IEEE Early Access Articles.

[59] N. Iotti, M. Picone, S. Cirani, and G. Ferrari, "Improving quality of experience in future wireless access networks through fog computing," *IEEE Internet Computing*, vol. 21, no. 2, pp. 26–33, Mar. 2017.

[60] Cisco. (Jan. 2014) Cisco delivers vision of fog computing to accelerate value from billions of connected devices. press release. [Online]. Available: http://newsroom.cisco.com/release/1334100/CiscoDeliversVision-of-Fog-Computing-to-Accelerate-Value-from-BillionsofConnected-Devices-utm-medium-rss

[61] M. Satyanarayanan, P. Bahl, R. Caceres, and N. Davies, "The case for vm-based cloudlets in mobile computing," *IEEE Pervasive Comput.*, vol. 8, no. 4, pp. 14–23, Oct. 2009.

[62] H. Li, G. Shou, Y. Hu, and Z. Guo, "Mobile edge computing: Progress and challenges," in *Proc. Fourth IEEE International Conf. on Mobile Cloud Computing, Services, and Engineering*, Oxford, UK, Mar. 2016, pp. 83–84.

[63] X. Chen, L. Jiao, L. Wenzhong, and F. Xiaoming, "Efficient multi-user computation offloading for mobile-edge cloud computing," *IEEE/ACM Trans. Netw.*, vol. 24, no. 5, pp. 2795–2808, Oct. 2016.

[64] ARCchart, "The mobile cloud: Market analysis and forecasts," 2011.

[65] C. Index, "Cisco visual networking index Forecast and methodology 2012–2017," *CISCO White Paper*, May 2013.

[66] G. Tyson, "A trace-driven analysis of caching in content-centric networks," in *Proc. IEEE 21st Int. Conf. Comput. Commun. Netw.*, Munich, Germany, Aug. 2012, pp. 1–7.

[67] J. Research. (Jul. 2011) Mobile cloud: Smart device strategies for enterprise & consumer markets 2011-2016. [Online]. Available: http://juniperresearch.com/

[68] M. Sheng, C. Xu, J. Liu, J. Song, X. Ma, and J. Li, "Enhancement for content delivery with proximity communications in caching enabled wireless networks: Architecture and challenges," *IEEE Commun. Mag.*, vol. 54, no. 8, pp. 70–76, Aug. 2016.

[69] J. Erman, A. Gerber, M. Hajiaghayi, D. Pei, S. Sen, and O. Spatscheck, "To cache or not to cache? the 3g case," *IEEE Internet Computing*, vol. 15, no. 2, pp. 27–34, Mar. 2011.

[70] S. Woo, E. Jeong, S. Park, J. Lee, S. Ihm, and K. Park, "Comparison of caching strategies in modern cellular backhaul networks," in *Proc. 11th annual international conference on mobile systems, applications, and services, ACM MobiSys*, Taipei, Taiwan, Jun. 2013, pp. 319–332.

[71] X. Wang, M. Chen, T. Taleb, A. Ksentini, and V. Leung, "Cache in the air: exploiting content caching and delivery techniques for 5g systems," *IEEE Commun. Mag.*, vol. 52, pp. 131–139, Feb. 2014.

[72] ETSI, "Network functions virtualisation (nfv): Use cases," *Tech. Rep. 2013001*, Mar. 2013.

[73] H. Ahlehagh and S. Dey, "Video caching in radio access network: Impact on delay and capacity," *in Proc. IEEE Wireless Communications and Networking Conference (WCNC)*, Shanghai, China, Jun. 2012, pp. 2276–2281.

[74] N. Golrezaei, K. Shanmugam, A. Dimakis, A. Molisch, and G. Caire, "Femtocaching: Wireless video content delivery through distributed caching helpers," in *Proc. IEEE INFOCOM12*, Mar. 2012, pp. 1107–1115.

[75] L. Han, B. Nath, L. Iftode, and S. Muthukrishnan, "Social butterfly: Social caches for distributed social networks," in *Proc. IEEE Third International Conference on Privacy, Security, Risk and Trust and IEEE Third International Conference on Social Computing*, Boston, MA, USA, Oct. 2011, pp. 81–86.

[76] L. Han, M. Puncevat, B. Nath, S. Muthukrishnan, and L. Ifte, "Social CDN: Caching techniques for distributed social networks," in *Proc. IEEE International Conference on Peer-to-Peer Computing (P2P)*, Tarragona, Spain, Sept. 2012, pp. 191–202.

[77] C. Zhang, Y. Sun, Y. Mo, Y. Zhang, and S. Bu, "Social-aware content downloading for fog radio access networks supported device-to-device communications," in *Proc. IEEE International Conference on Ubiquitous Wireless Broadband (ICUWB)*, Nanjing, China, Oct. 2016, pp. 1–4.

[78] O. Semiari, W. Saad, S. Valentin, M. Bennis, and H. Poor, "Context aware small cell networks: How social metrics improve wireless resource allocation," *IEEE Trans. Wireless Commun.*, vol. 14, no. 11, pp. 5927–5940, Nov. 2015.

[79] E. Zeydan, E. Bastug, M. Bennis, M. Kader, I. Karatepe, A. Er, and M. Debbah, "Big data caching for networking: Moving from cloud to edge," *IEEE Commun. Mag.*, vol. 54, no. 9, pp. 36–42, Sept. 2016.

[80] J. Tang and T. Quek, "The role of cloud computing in content-centric mobile networking," *IEEE Commun. Mag.*, vol. 54, no. 8, pp. 52–59, Aug. 2016.

[81] J. Tang, W. Tay, and T. Quek, "Cross-layer resource allocation with elastic service scaling in cloud radio access network," *IEEE Trans. Wireless Commun.*, vol. 14, no. 9, pp. 5068–5081, Sept. 2015.

[82] R. Tandon and O. Simeone, "Harnessing cloud and edge synergies: Toward an information theory of fog radio access networks," *IEEE Commun. Mag.*, vol. 54, no. 8, pp. 44–50, Aug. 2016.

[83] Q. Li, H. Niu, A. Papathanassiou, and G. Wu, "Edge cloud and underlay networks: Empowering 5g cell-less wireless architecture," in *Proc. European Wireless Conference*, May 2014, pp. 1–6.

[84] S. Park, O. Simeone, and S. Shitz, "Joint optimization of cloud and edge processing for fog radio access networks," *IEEE Trans. Wireless Commun.*, vol. 15, no. 11, pp. 7621–7632, Sept. 2016.

[85] J. Khan and Y. Ghamri-Doudane, "Saving: Socially aware vehicular information-centric networking," *IEEE Commun. Mag.*, pp. 100–107, Aug. 2016.

Chapter 2

Performance Metrics and Enabling Technologies

In this chapter, we present the performance metrics of and enabling technologies for integrated networking, caching and computing systems.

2.1 Performance metrics

We first present the general metrics of the entire integrated system, then we present the networking, caching and computing related metrics, respectively. All the metrics are classified and briefly described in Table 2.1.

2.1.1 General metrics

2.1.1.1 Cost

The cost metric refers to the investment paid by infrastructure providers and system operators for infrastructure construction and system operation, including CapEx and OpEx. For networking-caching-computing integrated systems, CapEX is composed of the expenses of constructing base stations, radio network controller equipment, backhaul transmission equipment, core network equipment, caching and computing equipment installation in network nodes. OpEx consists of energy charge, equipment lease and operation and maintenance fees [1]. The deployment cost includes the lease paid for renting licensed spectra issued by authorities.

Table 2.1: Performance Metrics of Networking-caching-computing Integrated Systems.

Types	Metrics	Units	Description
General Metrics	Cost	Currency units (e.g., US$)	Including CapEx and OpEx
	Revenue	Currency units (e.g., US$)	Earned revenue
	Recovery time	*second*	Time duration in which system recovers from failures
Networking-related Metrics	Coverage	m^3	A certain 3D area covered by wireless services
	Capacity (Throughput)	*bps*	Peak data rate for a certain area
	Deployment efficiency	$m^3/\$$ $Mbps/\$$	Throughput (capacity) / Deployment Costs
	Spectral efficiency	bps/Hz	Throughput (capacity) / bandwidth, in a certain coverage area
	Energy efficiency	bps/J	Throughput (capacity) / energy consumption
	QoS	various	Quality of service experienced by end users
	Service latency & Signaling delay	*second*	Packet delay and signaling delay
Caching-related Metrics	Average latency	*second*	Average content-delivery latency experienced by end users
	Hop-count	unitless	Number of network nodes traversed by a content request before it is satisfied
	Load fairness	unitless	Number of content requests received by one node / total number of content requests

Table 2.1: Performance Metrics of Networking-caching-computing Integrated Systems. (*continued*)

Types	Metrics	Units	Description
	Responses per request	unitless	Number of content responses received for one content request
	Cache hits	unitless	Number of content requests responded to by one network node
	Caching efficiency	unitless	Number of requested contents / total number of stored contents
	Caching frequency	unitless	Number of cached contents / number of contents flowing through the node
	Cache diversity	unitless	Number of distinctive contents cached in the network
	Cache redundancy	unitless	Number of identical contents cached in the network
	Absorption time	*second*	Time period that content remains in the node
Computing-related Metrics	Execution time	*second*	Time consumed by the execution of the computation task
	Energy consumption	*J*	Energy consumed for the execution of the computation task
	Computation dropping cost	various	Penalty associated with failure to accomplish computation tasks
	Throughput	*bps*	Data units delivered through the network per unit time interval

2.1.1.2 Revenue

Profit, which is defined as the difference between revenue and cost, can be used to reflect the effectiveness of networking-caching-computing integrated systems. Another indicator, revenue to cost ratio (RCR) can also be used.

2.1.1.3 Recovery time

The recovery time metric is an indicator of the resilience of the networking-caching-computing integrated system, which refers to the time duration in which the system recovers from a certain type of failure or attack [2].

2.1.2 Networking-related metrics

2.1.2.1 Coverage and capacity (throughput)

The coverage metric refers to the entire geographical area that is covered by the radio access network service, through which users can gain access to communication, caching and computing services provided by the integrated system. Capacity refers to the maximal aggregated peak rate (maximum theoretical throughput) of a certain territory served by a base station or access point [3]. Throughput refers to the aggregated data rate of the radio access network in a certain period of time and area. Coverage and capacity are the fundamental and crucial metrics in radio access network design and optimization, and are typically determined by bandwidth, transmitting power and network planning, which are succinctly described as follows.

Bandwidth

Bandwidth is the whole radio spectrum in a certain area that is available to mobile devices for accessing the communication, caching and computing services of the integrated system.

Transmitting power

Transmitting power is the power with which base stations and mobile devices transmit signals to each other.

Network planning

Network planning refers to topological design, synthesis and realization of the network, aiming at ensuring that the newly designed network or service can fully accommodate the demands of network operators and end users.

2.1.2.2 Deployment efficiency

The deployment efficiency (DE) metric is defined as the ratio of system throughput or coverage to deployment costs, which include both CapEx and

OpEx [1]. Typically, the DE metric is taken as an important network performance reference when conducting network planning.

2.1.2.3 Spectral efficiency

The spectral efficiency (SE) metric is defined as the fraction of system throughput or coverage to bandwidth. SE is usually taken as an important criterion in wireless network optimization [1]. Furthermore, more detailed SE metrics can further facilitate evaluating more specific network performances, for instance, the worst 5% users spectral efficiency and the cell edge spectral efficiency.

2.1.2.4 Energy efficiency

The energy efficiency (EE) metric is defined as the ratio of system throughput or coverage to energy consumption. It is worth noting that energy consumption includes not only signal transmission power consumption but also any other aspects of energy consumption in the entire network, including equipment energy and accessories, etc. [4].

2.1.2.5 QoS

The quality of service (QoS) metric is generally represented by several variables that characterize the network performance experienced by end users [5]. For example, in 3GPP LTE [6], the QoS metric is classified into 9 types of so-called QoS class identifiers (QCIs), which are associated to priority (9 levels), resource types (guaranteed bit rate or non-guaranteed bit rate), packet loss rate and packet delay budget.

2.1.2.6 Signaling delay and service latency

The signaling delay metric refers to the propagation delay of control signals between network entities that perform network management functions.

The service latency metric refers to the delay induced by preparation and propagation of data packets in the system. Since this chapter only focuses on radio access networks in the networking part, the technologies discussed in this chapter only involve three types of latencies that are specified as follows.

■ Propagation Delay

As the primary source of latency, propagation delay is defined as a function of how long it takes information to travel at the speed of light in wireless channels from origin to destination.

■ Serialization Delay

Serialization is the conversion of bytes (8 bits) of data stored in a device's memory into a serial bit stream to be transmitted over the wireless channels. Serialization takes a finite amount of time and is calculated as Serialization delay = packet size in bits / transmission rate in bits per second.

■ Queuing Delay

Queuing delay refers to the amount of time a data packet spends sitting in a queue awaiting transmission due to over-utilization of the outgoing link.

2.1.3 Caching-related metrics

2.1.3.1 Average latency

In caching-networking systems, the average content-delivery latency experienced by end users is a major performance metric. Typically, there are three types of average latency considered in the literature [7]. The first type is the average latency of delivering content to end users from clients' closest network nodes (such as routers) that cache the requested content locally. The second type is the average latency of delivering content to end users from peer network nodes. In this case, the requested content is not cached in the network node that is directly connected to the client but can be fetched from a peer network node in the same administrative domain. Note that this average latency includes the latency of content transmission from a peer node to a client associating node, and the latency of transmission from an associating node to a client. The third type is the average latency of delivering content from the content origin. These three types of average latency collectively reflect the average latency incurred by content delivery in the network. The latency metric is also referred to as *delay, download time* or *Round-Trip Time (RTT)*.

2.1.3.2 Hop-count

The hop-count is the number of network nodes that are traversed (inspected) by a content request before it is satisfied. In order to reflect the network traffic reduction brought by caching, the hop-count metric is always determined as a fraction, given the name *hop-count ratio, hop-reduction ratio* or *hop-count rate*, and is calculated as the ratio of the number of traversed nodes with caching being conducted to the number of traversed nodes without caching being conducted. This metric can be considered one of the estimations of the delay metric in the network [8].

2.1.3.3 Load fairness

The load fairness metric is defined as a ratio of the number of content requests and content responses received by one node to the total number of content requests and content responses generated in the whole network. This metric is intended to indicate the distribution of traffic and the existence of overloaded nodes.

2.1.3.4   Responses per request

This metric stands for the number of content responses being received by the content requester for one content request. This metric is intended to estimate the network traffic introduced by the multi-path mechanism.

2.1.3.5   Cache hits

The cache hits metric indicates the load of a node by measuring the number of content requests responded to (satisfied) by the node. When intended to reflect the load savings of a node due to conducting caching, the node hits metric is determined as a fraction called *cache-hit ratio*, which is normalized by the total number of content requests generated in the whole network. This server-hit ratio is also referred to as *cache-hit probability* or *cache-hit rate*.

2.1.3.6   Caching efficiency

Caching efficiency is defined as the fraction of the number of cached contents that have been requested during a certain period of time in a node to the total number of contents being stored in the same node.

2.1.3.7   Caching frequency

Caching frequency is defined as the fraction of the number of cached contents in a node to the total number of contents that have flowed through the same node during a certain period of time.

2.1.3.8   Cache diversity

The cache diversity metric refers to the number of distinctive contents cached in the whole network.

2.1.3.9   Cache redundancy

The cache redundancy metric refers to the number of identical contents cached in the whole network. Cache redundancy and cache diversity are the complements to one another.

2.1.3.10   Absorption time

The absorption time metric refers to the time period during which the cached content remains in the node. This metric is determined by cache replacement policies and content popularity distribution. It is also referred to as *caching time*.

2.1.4   Computing-related metrics

2.1.4.1   Execution time

The execution time metric refers to the time consumed by the execution of the computation task [9, 10, 11, 12, 13]. In other words, this metric represents the time period during which the computation is executed by network nodes, either user devices or servers. In mobile cloud computing and mobile edge computing systems, the time during which the computation task is transmitted to the MCC or MEC server may also be involved in this execution time metric [14, 15]. The data transmission time depends on network conditions such as bandwidth and interference level, instead of computing capabilities of network nodes. The execution time metric is also referred to as *execution latency* or *response latency* [16].

2.1.4.2 Energy consumption

The energy consumption metric refers to the energy consumed by network nodes, either user devices or servers, for the execution of the computation task. The energy consumption metric is in close relation to the execution time metric, since energy consumption depends on the time period during which the network nodes work. In MCC and MEC systems, the energy consumed by user devices for computation task offloading, and the energy consumed by the MCC or MEC servers for computation results feedback, can also be included in this energy consumption metric [14, 15].

2.1.4.3 Computation dropping cost

In some cases, for various reasons, the computation task cannot be executed but has to be dropped. For instance, the task is too computationally intensive to be executed on a mobile device, while the MEC server is already overloaded or the wireless link between a mobile device and the MEC server is undergoing deep fading. In this case, users' quality of experience (QoE) can deteriorate due to the computation request not being satisfied. Therefore the computation dropping cost is a penalty associated with this failure [17].

2.1.4.4 Throughput

The throughput metric refers to the data units delivered through the network per unit time interval. This metric is often used in mobile cloud computing and mobile edge computing systems to serve as a joint indicator of the network transmission capability and the network nodes computation capability [18].

2.2   Enabling technologies

In this section, a number of the enabling technologies for integrated networking, caching and computing systems are summarized. These technologies are presented in taxonomy based on the types of integrations.

2.2.1 Caching-networking

2.2.1.1 Caching in heterogeneous networks

As a key component of 5G cellular communication systems, heterogeneous networking has been extensively studied, and caching approaches and corresponding caching gains are well investigated. A number of typical caching schemes and supporting mathematical tools are introduced and discussed here.

A dynamic caching paradigm in hyper-dense small cell networks is proposed in [19] to address the caching issue in realistic scenarios. This paper poses the problem that the scenarios considered in the literature are always in a static environment, in which mobile users are assumed to remain stationary during data transmitting processes so that the downloading procedure can always be completed by associated small cell base stations. In order to address the caching problem more realistically, the proposed caching paradigm emphasizes a dynamic environment in which mobile devices connect to multiple small cell base stations intermittently; therefore the mobile devices may download only part of the cached data due to limited connecting durations. The failed transmitting tasks are then redirected to a macro cell base station. Aiming at minimizing the load of the macro cell, this work uses random walks on a Markov chain to model mobile user movements in an optimization framework. The proposed caching paradigm leverages the combination of user mobility predictions and information-mixing methods based on the principle of network coding.

An algorithm is proposed in [20] to jointly consider caching, routing, and channel assignment for video delivery in coordinated small-cell wireless cellular systems of the future Internet, aiming at maximizing the throughput of the system. This work stresses three major issues. First, collaborations between small cells are taken into account in cached content delivery. When a content request cannot be satisfied by the user's associated small cell base station, instead of redirecting the request to a macro cell base station immediately, the request is forwarded to other small cell base stations to search for possible satisfaction. Second, the wireless transmission interference between cache-to-user communication and other links transmitting is taken into consideration to optimize throughput. Third, cached content delivery routings are handled differentially depending on the geographical locations of the mobile users. This work formulates the above described issues into a linear program. The column generation method is applied to solve the problem by splitting it into subproblems, and an approximation algorithm is utilized to further reduce the complexity of the problem.

The authors of [21] consider cooperative caching and content delivery in a heterogeneous network, in which the spectrum is shared between one macro cell and a number of small cells. Cooperation among cells is realized by the operation in which small cell base stations cache primary files that are requested by macro cell users, so that some of the content requests of the macro cell users can be satisfied by small cell base stations. Two caching and delivery algorithms are developed in this cooperative caching scenario. In the first algorithm the priority of access of the primary files is the same as that of secondary files (files that are for small cell users), while in the second algorithm the priority of the primary files is higher than that of secondary files. Due to a cooperation mechanism, both of the algorithms can maintain a greater set of supportable request generation rates than that of non-cooperative solutions.

It is conspicuous and unanimously agreed that caching in small cell networks is beneficial to relief of backhaul and improvement of throughput, but it remains ambiguous whether caching is helpful to the improvement of energy efficiency (EE) of the network. Therefore, the work in [22] explores the impact of caching on EE in small cell networks. The closed-form expression of the approximated EE is derived. Based on that, the conditions under which the EE can benefit from caching are given, and the optimal cache capacity that can maximize the EE is deduced. Furthermore, the maximal EE gain brought about by caching is analyzed accordingly. The major observation of their exploration is that caching at small cell base stations can benefit the network EE with power-efficient cache hardware employed, and that if local caching already provided EE gains, caching more content may not further benefit the network EE.

2.2.1.2 Caching in information-centric networking

In order to address the dramatic growth of mobile data traffic and the shift of mobile users' service preference from peer-to-peer data transmission to information dissemination oriented services, information-centric networking has risen to serve as a promising solution.

A collaborative in-network caching solution with content-space partitioning and hash-routing in information-centric networking is studied in [23]. By partitioning the content space and assigning partitions to caches, the path stretch incurred by hash-routing can be constrained by the proposed scheme. The issue of assigning partitions to caches is formulated as an optimization problem aiming at maximizing the overall hit ratio, and a heuristic algorithm is developed to solve the optimization problem. Furthermore, the partitioning proportion issue is formulated as a min-max linear optimization problem, in order to balance cache workloads.

The authors of [24] focus on encrypted in-network caching in information-centric networking. In order to address the newly raised user privacy exposure and unauthorized video access issues, this work presents a compact and encrypted video fingerprint index to enable the network to locate the cached encrypted chunks for given encrypted requests. Furthermore, for video delivery purposes, a secure redundancy elimination protocol which leverages the cached encrypted chunks is designed.

In [25], a dynamic approach is proposed in information-centric cognitive radio networks (IC-CRNs) to improve the performance of video dissemination, by leveraging harvested bands for proactively caching video contents. The harvested bands allocation issue is formulated as a Markov decision process with hidden and dynamic parameters, which is further transformed into a partially observable Markov decision process and a multi-armed bandit formulation. A spectrum management mechanism is designed accordingly to maximize the revenue of proactive video caching and the efficiency of spectrum utilization.

The authors of [7] investigate the problem of tradeoffs between in-network storage provisioning costs and network performance in content-centric networks. A holistic model is formed to characterize the cost of global coordination of in-network storage capability and network performance of delivering content to users. The strategy for provisioning the storage capability that jointly optimizes the network cost and performance is derived accordingly.

2.2.1.3 Caching in D2D networking

As an important component of 5G wireless cellular networks, device-to-device (D2D) communication has attracted extensive attention from researchers. A major functionality of D2D communication is content sharing between mobile devices, and therefore caching strategy is essential in D2D techniques.

The authors of [26] propose a content caching scheme in D2D networking to maximize the probability of successful content delivery, which is derived by using stochastic geometry in the presence of interference and noise. An optimization problem is formulated to search for the caching distribution that can maximize the density of successful receptions (DSR). Content requests are modeled as a Zipf distribution

In [27], the authors also study caching schemes in D2D networking. Unlike the work in [26], power allocation is taken into consideration in this work. An optimization problem in which D2D link scheduling and power allocation are jointly considered is formulated, aiming at maximizing system throughput. Furthermore, the problem is decomposed into an optimal power allocation problem and a D2D link-scheduling problem, in order to tackle the nonconvexity of the original problem. The D2D link-scheduling problem is solved by an algorithm maximizing the number of D2D links under the constraints of the transmitting power and the signal to interference plus noise ratio, while the power allocation problem is solved by another algorithm maximizing the minimum transmission rate of the scheduled D2D links.

Unlike the works in [26] and [27], which solely focus on D2D caching schemes, the work in [28] studies the issue of jointly caching and transmitting content in both small cell networking and D2D networking. This work considers two types of caching gains, namely, pre-downloading and local caching gains, and these two types of gains have an inherent tradeoff. In this context, a continuous time optimization problem is formulated, aiming at the determination of optimal transmission and caching policies that minimize a generic cost function, such as throughput, energy and bandwidth. It is then demonstrated that caching files at a constant rate offers the optimal solution, due to the fact that it allows reformulation of the original problem as a finite-dimensional convex program.

2.2.1.4 Others

Some approaches that specify other network architectures are introduced here.

The authors of [29] study caching-assisted video streaming in multicell multiple-input multiple-output (MIMO) networks. A proportion of the video files are cached in base stations and these cached files opportunistically induce Coop-MIMO transmission, which can achieve a MIMO cooperation gain without expensive payload backhaul. In this context, the approximated video streaming performance is derived by using a mixed fluid-diffusion limit for the playback buffer queueing system. Based on the performance analysis, the cache control and playback buffer management issues are formulated as a stochastic optimization problem. Then a stochastic subgradient algorithm is employed to obtain the optimal cache control, and a closed form solution for the playback buffer thresholds is given.

Since the economic viability of cellular networks is being challenged by the dramatically increasing mobile data, the work in [30] studies the solution of leasing wireless bandwidth and cache space of access points (APs) in residential 802.11 (WiFi) networks, in order to reduce cellular network congestion and relieve the backhaul link payload of the APs. Monetary incentive (compensation) is provided to APs, in proportion to the amount of bandwidth and cache space that they are leasing, which in turn influences caching and routing policies. In this context, a framework taking into account joint optimization of incentive, caching, and routing policies is proposed.

The authors in [31] study the success rate of cached content delivery in wireless caching helper networks, and draw the conclusion that the success rate mainly depends on cache-based channel selection diversity and network interference. They further point out that with given channel fading and network topology, both interference and channel selection diversity vary corresponding to the caching strategies of the caching helpers. In order to maximize the average success probability of content delivery, the paper studies probabilistic content placement and derives optimal caching probabilities in the closed form using stochastic geometry, to desirably control channel selection diversity and network interference.

2.2.2 Computing-networking

2.2.2.1 Cloud computing and networking

Mobile cloud computing (MCC) is a very typical paradigm of the integration of computing and networking functionalities. By allowing mobile users to offload their computation tasks to a remote cloud, which is typically located in the Internet, MCC endows excessive computational capabilities on wireless communication networks, enabling the execution of computationally intensive mobile applications such as natural language processing and augmented reality [1].

In order to leverage the intensive computation resource of mobile cloud computing on the processing of data gathered by ubiquitous wireless sensor networks (WSNs), some researchers focus on the integration of these two network paradigms. In [33], a framework concerning the processing of the sensory data in WSN–MCC integration is proposed. The framework aims at delivering desirable sensory data to mobile users fast, reliably and securely, prolonging the WSN lifetime, reducing storage requirements, and decreasing traffic load. Moreover, the framework can monitor and predict the future trend of sensory data traffic. In [34], an integrated WSN–MCC architecture is proposed, in which the cloud is considered as a virtual sink with a number of sink points, and each of the sink points gathers the data from sensors within a zone. Thus, the sensor data is eventually stored and processed in a decentralized manner in the cloud. The work in [35] considers collaborative location-based sleep scheduling schemes in an integrated network of MCC and WSN. In order to address mobile users' current location dependent specific data requests and decrease energy consumption of sensors, the proposed scheme dynamically determines the awake or asleep status of sensors based on the current locations of mobile users.

Collaborative task execution between mobile devices and mobile cloud servers relies on the quality of wireless communication provided by wireless networks. The research of [36] investigates this issue under a stochastic wireless channel. The task offloading and collaborative task execution strategies are formulated as a constrained shorted path problem, aiming at minimizing the energy consumption of mobile devices. A one-climb policy and an enumeration algorithm are proposed to solve the problem. Furthermore, a Lagrangian relaxation based aggregated cost algorithm is employed to approximately solve the problem with lower complexity.

A distributed function computation in a noisy multihop wireless network scheme is considered in [37]. An adversarial noise model is adopted as the channel model, and only divisible functions are considered. A general protocol for evaluating any divisible functions is designed and the bottleneck of the protocol is analyzed to offer insights into design of more efficient protocols.

The authors of [38] propose a solution for the integration of mobile cloud computing and microwave power transfer (MPT), to offer computation services to passive low-complexity devices such as sensors and wearable computing devices. Mobile users can offload their computation through BSs to a cloud or execute them locally, and meanwhile, the BSs can transfer power to mobile users. A set of policies for computation offloading decisions and MPT time divisions is proposed, aiming at minimizing energy consumption and maximizing harvested energy. The problem is solved by convex optimization theory.

The authors of [11] focus on energy saving in mobile cloud computing systems with a stochastic wireless channel. In the local computing case, the CPU frequency is reconfigured, and in the offloading case, the data transmission rate is dynamically adjusted according to the stochastic channel condition, both aiming at minimizing mobile device energy consumption. The scheduling policies are formulated as constrained optimization problems, and closed-form solutions are obtained for the problems.

In [33], a solution for jointly considering wireless networking and cloud computing is proposed to improve the end-to-end performances of cloud mobile media delivery. The spectral efficiency in wireless network and pricing information in the cloud are formulated as a Stackelberg game model, in which the power allocation and interference management are performed on the basis of pricing of media services. The replicator dynamics method is adopted to solve this game.

The work in [40] proposes a resource allocation scheme in cognitive wireless networks with cloud computing. A virtualized data center model is derived to handle the dynamic resource allocation problem under fluctuating channel and workload conditions, aiming at minimizing energy consumption and maximizing long-term average throughput while maintaining network stability.

The authors of [41] consider wireless resource scheduling in multiuser multiservice mobile cloud computing systems with different priorities for real-time services and non-real-time services. An M/M/1 queueing model is formulated to improve the quality-of-service (QoS) of real-time streams and minimize the average queueing delay of non-real-time services. A queueing delay-optimal control algorithm and a delay-constrained control algorithm are developed, respectively, to solve the problems.

In [42], the authors investigate security issues in a virtualized cloud computing paradigm–Amazon Elastic Compute Cloud (EC2). They demonstrate that in EC2, an adversary can deliberately delay the execution of a targeted application by utilizing the competition for shared resources between virtual I/O workloads. A framework called Swiper is designed and presented as a tool to slow down the targeted application and virtual machine with minor resource consumption.

2.2.2.2 Fog computing and networking

In order to reduce the transmission latency of cloud computing systems and provide localized and location-based information services, fog computing is proposed as a complement of cloud computing, bringing computation resources into proximity to end users. In essence, fog computing serves as an intermediate layer between the cloud and mobile devices.

The authors of [43] study the cooperation and interplay between the fog and the cloud. The workload allocation between the fog and the cloud is formulated as an optimization problem, aiming at minimizing power consumption under the service delay constraint. An approximate algorithm which decomposes the primal problem into several subproblems is used to solve the original problem.

In [44], a new framework called device-to-device (D2D) fogging is proposed to enable the sharing of computation and communication resources among mobile devices. The framework is enabled by D2D collaboration under the assistance of a network control. An optimization problem is formulated in this framework to minimize the time-average energy consumption of task execution of all mobile devices, taking into consideration incentive constraints to prevent over-exploiting and free-riding behaviors. An online task offloading algorithm that leverages Lyapunov optimization methods is derived accordingly.

Similar to [44], the work in [45] presents a distributed architecture in a vehicular communication network, called vehicular fog computing, in which vehicle end users are utilized as infrastructures to collaboratively provide computation and communication services to the whole network.

The authors of [46] study the energy consumption issue of nano data centers (nDCs) in fog computing systems. Nano data centers are tiny computers that are distributed in end user premises, and they serve as Internet of Things servers to handle content and applications in a peer-to-peer (P2P) fashion. Flow-based and time-based models are employed to investigate the energy consumption of shared and unshared network equipment, respectively.

In order to provide sufficient computation resources for video or image data processing in incidental disasters, the work of [47] proposes an incident-supporting visual cloud computing solution, bringing computation resources from the cloud to the proximity of the incident sites. The proposed collection, computation and consumption solution leverages SDN to construct a fog-cloud computing integrated architecture, in which fog computing is deployed at the network edge, close to data collection/consumption sites, and coupled with a cloud computing core, which provides major computation resources via a computation offloading operation.

The authors of [48] study resource management and task scheduling issues in a fog computing supported software-defined embedded system, in which computation tasks can be executed by either embedded devices or a fog computing server. The resource allocation, the I/O interrupt requests balance issues and the task scheduling between mobile devices and fog computing servers are formulated into one mixed-integer nonlinear programming problem to search for effective solutions.

In order to manage the physical diversity of devices and bridge different networks, [49] proposes a type of special fog node, called an IoT Hub. An IoT Hub is placed at the edge of multiple networks, offering the services of border router, cross-proxy, cache, and resource directory.

Due to the distributed nature of fog nodes and uncertainty on arrival of fog nodes within the network, dynamic and optimal fog network formation and task distribution is a problem that needs to be solved. The work in [50] is dedicated to this topic. An approach based on the online secretary framework is proposed. In the framework, any given task-initiating fog node is able to choose its desired set of neighboring fog nodes and perform task offloading, aiming at minimizing the maximum computational latency.

2.2.2.3 Mobile edge computing and networking

As a recently emerging computing-networking paradigm, mobile edge computing (MEC) aims at providing wireless communication networks with low-latency computation services, by allowing mobile users to offload their computationally intensive workloads to MEC servers placed in proximity to end users. The major issues in MEC systems are computation offloading decisions and resource allocation.

Reference [14] simply focuses on the offloading decision problem in MEC systems. The investigation of the paper argues that the multi-user computation offloading problem in a multi-channel wireless interference environment is an NP-hard problem, so the problem is formulated into a multi-user computation offloading game, and further investigation on the game property shows that the game admits a Nash equilibrium. Therefore, a distributed offloading algorithm that achieves a Nash equilibrium is designed accordingly to solve the game.

In order to reduce the frequency of interrupted computation tasks due to battery depletion and achieve green computing, [17] investigates a green MEC system with energy harvesting technology and proposes a computation offloading and resource allocation solution in this system, taking into account the offloading decision, the CPU-cycle frequencies for mobile execution, and the transmitting power for computation offloading. Aiming at minimizing the execution cost, which addresses execution latency and task failure, a Lyapunov optimization-based dynamic computation offloading algorithm is presented for problem solving.

The authors of [12] consider the joint optimization of radio resources (the transmit precoding matrices of mobile devices) and computation resources (the assigned CPU cycles/second) during MEC operations in a MIMO multicell system. This issue is formulated as an optimization problem, aiming at minimizing the overall users' energy consumption, while meeting latency constraints. The closed-form global optimal solution is derived in the single-user case, and an iterative algorithm utilizing the successive convex approximation technique is applied to solve the problem in a multiuser scenario.

Unlike the studies in [14] and [17], the work in [51] considers a partial computation offloading approach, in which only part of the computation task of a UE is offloaded to the MEC server, while the rest of it is executed locally in a device. Two optimization problems are designed with objectives of energy consumption of UE minimization (ECM) and latency of application execution minimization (LM), respectively. The computational speed of UE, transmit power of UE and offloading ratio are adopted as variables in these two optimization problems. The non-convex ECM problem is transformed into a convex problem by using a variable substitution technique and solved, and the LM problem is solved by a locally optimal algorithm with the univariate search technique.

We propose an integrated framework for interference management and computation offloading in wireless networks with MEC in [52, 53]. In the framework, the computation offloading decision, physical resource block allocation and MEC computation resource allocation are formulated as an optimization problem. The computation offloading decisions are made by the

MEC server according to load estimations, then the PRBs are allocated by a graph coloring method. Finally, the outcomes of PRB allocation and offloading decisions are used to distribute computation resources to UE.

2.2.3 Caching-computing-networking

Reference [54] considers the data prefetching issue in wireless computing over fluctuating channel conditions. Mobile users can prefetch data for computing from access points when the channel condition is good, in anticipation of requesting these data during bad quality channel periods. The issue is whether to prefetch or not in each time point, taking into consideration memory constraints, application latency and channel fluctuations. A dynamic programming approach is designed accordingly to derive the optimal prefetching policy.

Nowadays, large-scale video distribution constitutes a considerably large proportion of Internet traffic. In [55], a solution utilizing Media Cloud is proposed to deliver on-demand adaptive video streaming services. The caching, transcoding and bandwidth costs at each edge server are formulated as an optimization problem, aiming at minimizing the total operational cost. Subsequently, the closed-form solution of optimal caching space allocation and transcoding configuration at each edge server is derived through a two-step approach.

As a typical application of integrated caching, computing and networking systems, video transcoding and caching received considerable attention in literature. Reference [56] proposes a partial transcoding scheme for content management in heterogenous networks with a media cloud. Video content is encoded and separated into segments, parts of which are stored, resulting in a caching cost, while others are transcoded online, resulting in a computing cost. In this context, a constrained stochastic optimization problem is formulated to determine whether a segment is cached or transcoded, aiming at minimizing the long-term overall cost. An algorithm utilizing a Lyapunov optimization framework and Lagrangian relaxation is designed to solve the problem.

Similar to [56], [57] considers the caching-computing tradeoff problem in a cloud computing environment. For each query, a web application can either choose to compute the response fresh, which causes a computing cost, or cache the response to reduce the future computing cost, which causes a caching cost. This problem is abstracted as a variant of the classical Ski-Rental problem. A randomized algorithm is proposed accordingly to provide arrivals-distribution-free performance guarantees.

A simulation framework is proposed in [58], to serve as an evaluation tool for transparent computing systems with particular cache schemes. The simulation framework enables the evaluation of the performance of multi-level cache hierarchies with various cache replacement policies in transparent computing systems.

Reference [59] utilizes a big data tool to investigate the proactive caching issue in 5G wireless networks, aiming at optimizing the caching gain in terms of request satisfactions and backhaul offloadings. This work proposes that content popularity estimation plays an important role in optimization of the caching gain. The users' mobile traffic data in the network is collected and analyzed by big data platforms deployed at base stations for content popularity estimation.

Reference [60] focuses on the study of moving data-intensive video distribution applications to a Cache A Replica On Modification (CAROM) cloud file system, aiming at reducing access latency while maintaining a low storage cost. A scheduling mechanism is proposed to serve as a lubricant between data-intensive applications and CAROM. In the proposed scheme, a tripartite graph is adopted to clarify the relationships among computation nodes, data nodes and tasks. A framework is designed for the situation in which the task data is stored in the cache, followed by the design of two variant frameworks considering the performance of task and limitations of cache size. Finally, a k-list algorithm is employed to serve as an approximation algorithm.

As an efficient tool for resource sharing, network virtualization has been widely applied to cloud computing systems. Virtualization approaches can consolidate application workloads into shared servers. Due to multiple computing cores, multi-level caching hierarchy and memory subsystems, there exist contentions between virtualized resources. Reference [61] proposes hardware platform specific counters to detect those contentions. Furthermore, a software probe based scheme is designed to enhance contention detection.

Reference [2] studies the decentralized coded content caching in a multihop device-to-device communication scenario of next generation cellular networks. The contents are linearly combined and stored in caches of user devices and femtocells, and the throughput capacity is compared to that of decentralized uncoded content caching. It is demonstrated that, due to content coding, the number of hops needed to deliver the requested content is reduced, and thereby the decentralized coded content caching can achieve a better network throughput capacity compared to decentralized uncoded caching. Moreover, it is shown that under Zipfian content request distribution, the decentralized coded content cache placement is able to increase the throughput of cellular networks by a factor of $(log(n))^2$, where n stands for the number of nodes served by a femtocache.

We formulate the computation offloading decision, content caching strategy and spectrum and computation resources allocation issues as an optimization problem in our previous papers [3, 4]. The original non-convex problem is transformed into a convex problem and further decomposed into a number of sub-problems. Then a distributed algorithm called the alternating direction method of multipliers (ADMM) is employed to solve the problem in a more efficient way compared to a centralized algorithm.

Finally, the enabling technologies discussed above are summarized in Table 2.2. Furthermore, we also provide a comparison of pros and cons of all the technologies in Table 2.3.

Table 2.2: Enabling technologies for integrated networking, caching and computing.

Integration	*Ref.*	*Networking*	*Purposes*	*Contributions*
	[19]	Heterogeneous network	Minimizing the load of macro cell	A dynamic caching scheme considering moving users
	[20]	Heterogeneous network	Maximizing throughput	Jointly addressing caching, routing, and channel assignment issues for video delivery
	[21]	Heterogeneous network	Maximizing the set of supportable request generation rates	Enabling cooperative caching between macro cell and small cells
	[22]	Heterogeneous network	Maximizing energy efficiency	Giving the closed-form expression of EE and the caching policy that maximizes EE
	[23]	Information-centric network	Maximizing the overall cache hit ratio	A collaborative caching solution with content-space partitioning and hash-routing
	[24]	Information-centric network	Solving the user privacy exposure and unauthorized video access issues	Presenting an encrypted video fingerprint index to enable the network to locate the cached encrypted chunks for given encrypted requests
	[25]	Information-centric network	Maximizing revenue of caching and spectral efficiency	Utilizing Markov decision process to allocate harvested bands
Caching-networking	[7]	Information-centric network	Tradeoff between storage provisioning cost and network performance	Providing the model of provisioning cost and performance, and jointly optimizing them
	[26]	D2D network	Maximizing the probability of successful content delivery	Leveraging stochastic geometry to search for the optimal caching distribution
	[27]	D2D network	Maximizing throughput	A caching scheme taking into account power allocation
	[28]	D2D network	Minimizing a generic cost function, such as throughput, energy and bandwidth	Jointly considering caching and transmitting content in both small cell networking and D2D networking
	[29]	MIMO network	Optimizing video streaming performance	Providing the optimal cache control and a closed form solution for the playback buffer thresholds
	[30]	IEEE 802.11 network	Relieving the backhaul link payload of APs	A framework jointly considering incentive, caching, and routing policies
	[31]	Wireless caching helper network	Maximizing success probability of content delivery	Utilizing stochastic geometry to derive optimal caching probabilities

Continued on next page

Table 2.2: Enabling technologies for integrated networking, caching and computing (continued).

Continued from previous page

Integration	Ref.	Networking	Purposes	Contributions
	[33]	Wireless sensor network	Reliable and secure sensory data delivery with low storage requirements	Presenting a sensory data processing framework that can monitor and predict the future trend of traffic
	[34]	Wireless sensor network	Fast processing of sensory data	Presenting cloud sink model
	[35]	Wireless sensor network	Decreasing energy consumption	Presenting a collaborative location-based sleep scheduling scheme
	[36]	Heterogeneous network	Minimizing energy consumption	Enabling collaborative task execution between mobile devices and cloud servers
	[37]	Heterogeneous network	Evaluating distributed function computation	Presenting a protocol for evaluating divisible functions
	[38]	MPT wireless network	Minimizing energy consumption and maximizing harvested energy	Enabling integration of MCC and MPT
	[11]	Heterogeneous network	Minimizing mobile device energy consumption	Presenting energy saving scheme under stochastic wireless channel
	[33]	Heterogeneous network	Improving end-to-end performances	Jointly considering wireless networking and cloud computing
Computing-networking	[40]	Cognitive wireless network	Minimizing energy consumption and maximizing throughput	Dynamic resource allocation scheme under fluctuating channel conditions
	[41]	Heterogeneous network	Improving QoS and minimizing queueing delay	Presenting a scheme for reducing service delay in multiuser multiservice MCC
	[42]	EC2	Investigate security issues	A framework for security peoblem investigation
	[43]	Heterogeneous network	Minimizing power consumption	Enabling cooperation and interplay between the fog and the cloud
	[44]	D2D network	Minimizing energy consumption of mobile devices	Enabling the sharing of computation and communication resources among mobile devices
	[45]	Vehicular communication network	Proposing the vehicular fog computing framework	Enabling the cooperation on computation and communication between vehicles
	[46]	Heterogeneous network	Minimizing energy consumption	Presenting the framework utilizing nano data centers in P2P pattern
	[47]	Software defined networking	Providing computation resources to incident-supporting data processing	Enabling the integration of fog computing and SDN to support disaster rescue
	[48]	Software defined networking	Improving end-to-end experience	Presenting an approach for resource management and task scheduling

Continued on next page

Table 2.2: Enabling technologies for integrated networking, caching and computing (continued).

Continued from previous page

Integration	Ref.	Networking	Purposes	Contributions
Computing-networking	[49]	Multiple networks	Bridging different networks	Managing the diversity of devices and networks
	[50]	Heterogeneous network	Minimizing the maximum computational latency	Solving the problems of fog network formation and task distribution
	[14]	Heterogeneous network	Optimal offloading solution	Utilizing game theory to solve multi-user offloading problem
	[17]	Heterogeneous network	Minimizing execution cost	Jointly addressing offloading and energy harvesting problems
	[12]	MIMO multicell network	Minimizing user energy consumption	Jointly considering radio and computation resources in a MIMO system
	[51]	Heterogeneous network	Minimizing energy consumption and latency	Enabling partial computation offloading
	[52]	Heterogeneous network	Minimizing overall system overhead	Enabling the joint optimization of computation offloading and interference management
Caching-computing-networking	[54]	Heterogeneous network	Enhancing user experience	Optimal data prefetching policy
	[55]	Heterogeneous network	Minimizing operational cost	Jointly addressing caching, transcoding and bandwidth issues in video distribution
	[56]	Heterogeneous network	Minimizing overall cost	Enabling partial transcoding and caching
	[57]	Heterogeneous network	Minimizing overall cost	Optimal tradeoff between caching and computing
	[58]	Heterogeneous network	Performance evaluation	A simulation framework for caching-computing-networking systems
	[59]	Heterogeneous network	Optimizing caching gain	Utilizing big data platform for caching estimation
	[60]	Heterogeneous network	Reducing access latency	Enabling video distribution applications in CAROM
	[61]	Virtualized network	Detecting contentions between virtualized resources	Enabling contention detection in virtualized MCC systems
	[2]	D2D network	Maximizing throughput	Enabling coded caching in 5G D2D networks
	[3]	Heterogeneous network	Maximizing system utility	Enabling joint optimization of computation offloading, resource allocation and caching strategy

Table 2.3: Advantages and shortcomings of the enabling technologies.

Integration	Ref.	Advantages	Shortcomings
Caching-networking	[19]	Optimization is conducted under a dynamic user movement model.	The workload fairness between macro cell and small cells is not taken into consideration.
	[20]	Small cell collaboration is leveraged in video content search delivery process.	The content search and transmitting latency is considerably large.
	[21]	Cooperation between macro cell and small cells is realized in cache replacement and delivery procedures.	The popularity of cached content is not considered.
	[22]	An analysis about the benefit of caching on energy efficiency (EE) is provided.	The energy efficiency model is a static one.
	[23]	Content space partitioning is utilized to maximize the hit ratio.	The optimization problem is solved by a heuristic algorithm.
	[24]	User privacy exposure and unauthorized video access issues are addressed by an encrypted caching approach.	Latency and throughput properties are not taken into account.
	[25]	The harvested bands allocation issue is addressed on a dynamic Markov decision process model.	The overall throughput optimization is not considered when conducting spectrum management.
	[7]	The coordination of storage provisioning cost and network performance is conducted from a holistic perspective.	The tradeoff between cost and performance is not conducted on a dynamic basis.
	[26]	The caching scheme is derived in the presence of interference and noise.	The content request model is static.
	[27]	Power allocation is considered when conducting caching scheduling.	The caching scheduling merely considers the maximization of the number of D2D links.
	[28]	Jointly considered the caching issues in both small cell networking and D2D networking.	The cooperation and combination of these two types of networking are not neatly elaborated.
	[29]	MIMO transmission is leveraged in the transmitting of cached content.	MIMO transmission can only be induced occasionally.
	[30]	Spectrum and cache space in WiFi networks is utilized for cellular network caching.	The resources leasing and caching scheduling model is static.
	[31]	An analysis on content delivery success rate is given.	The analysis under fluctuating channel fading is not given.

Continued on next page

Table 2.3: Advantages and shortcomings of the enabling technologies (continued).

Continued from previous page

Integration	Ref.	Advantages	Shortcomings
	[33]	Future trend of sensory data traffic can be predicted by utilizing MCC.	Energy efficiency of the data processing scheme is not considered.
	[34]	Data gathering and processing workload is distributed in the system, thus alleviating congestion.	The workload fairness among different areas of the cloud is not discussed.
	[35]	The mobile users' current location dependent specific data requests are considered.	The potential loss due to sensors sleeping is not analyzed.
	[36]	Collaboration between mobile devices and cloud servers is realized.	The control of energy consumption of the cloud server is not considered.
	[37]	An evaluation on divisible functions in multihop wireless network is given.	The protocol is on a basis of constant data size.
	[38]	Microwave power transfer is integrated with computing systems.	Computation workload balancing is not considered when conducting computation offloading decisions.
	[11]	A stochastic wireless channel model is adopted.	In offloading case, the CUP frequency is not optimally reconfigured.
	[33]	The power allocation and interference management are jointly considered in computing and communication scheduling.	The pricing of media services is simple and arbitrary.
Computing-networking	[40]	A dynamic resource allocation problem is addressed under fluctuating channel and workload conditions.	Network latency is not considered.
	[41]	Different priorities of real-time and non-real-time services are taken into account.	The optimal tradeoff of delay between these two types of services is not analyzed.
	[42]	The security issue of computing systems is investigated on a specific use case.	The conclusion is drawn on a specific use case and hence lacks generality.
	[43]	The advantages of fog and cloud computing are combined with each other.	The minimum throughput constraint is not considered in the optimization.

Continued on next page

Table 2.3: Advantages and shortcomings of the enabling technologies (continued).

Integration	Ref.	Advantages	Shortcomings
Continued from previous page			
	[44]	Computation resources of mobile devices can be shared through collaboration of devices.	The network control factor is not explicitly formulated in the optimization problem.
	[45]	Computation and communication resources are shared among vehicles in a certain area.	Workload fairness between vehicles is not considered.
	[46]	Nano data centers are utilized to complement and reinforce the computation capability of fog computing systems.	Computation execution time consumption is not taken into account.
	[47]	The scheme is designed specifically for incident-supporting purposes.	The independent operation of fog computing and the workload allocation between fog and cloud are not sufficiently discussed.
	[48]	The computation scheduling approach is designed specifically for software-defined embedded systems.	The latency of computation execution is not discussed.
	[49]	The same kind of questions have rarely been addressed before.	ack of mathematical analysis.
	[50]	The same kind of questions have rarely been addressed before.	It is hard to guarantee a globally optimal solution and lacks discussion on this problem.

Continued on next page

Table 2.3: Advantages and shortcomings of the enabling technologies (continued).

Continued from previous page

Integration	Ref.	Advantages	Shortcomings
Computing-networking	[14]	The optimal offloading decision can meet a Nash equilibrium of the formulated game.	Solely focuses on offloading decision without optimization of resources allocation.
	[17]	Energy can be conserved while performing computation offloading.	The probability of computation task interruption is not considered in the optimization problem.
	[12]	Radio and computation resources are jointly considered in a MIMO environment.	The fairness of energy consumption among different UE is not considered.
	[51]	Partial computation offloading makes the system more agile and more flexible.	The slicing of computation tasks is not practical.
	[52]	The mutual interference among different offloading procedures is mitigated.	The method is heuristic.
Caching-computing-networking	[54]	A dynamic programming approach is designed over fluctuating channel conditions.	Only data prefetching issue is addressed, and no other problem such as resource allocation is jointly considered.
	[55]	Closed-form solution for caching space allocation and transcoding configuration is derived.	The latency performance, which is essential to video services, is not taken into consideration.
	[56]	The optimal tradeoff between video caching and transcoding is derived.	The user experience, which is crucial to video services, is not considered.
	[57]	The optimal tradeoff between video caching and transcoding with arrivals-distribution-free performance guarantee is given.	atency and user experience factors are not considered.
	[58]	A simulation framework enabling multiple cache replacement policies is described in detail.	The framework cannot be applied to other computing systems.
	[59]	Big data platform gives more accurate and more reliable cache popularity estimation.	Big data is merely leveraged on caching estimation, and is not used for resource allocation.
	[60]	Video access latency is reduced while maintaining low caching cost.	The study is conducted exclusively on a CAROM system and hence lacks generality.
	[61]	One of the few works that focus on contention detection in virtualized networks.	Contention prevention or contention alleviation schemes are not discussed.
	[2]	A pioneering piece of work on coded content caching in D2D networks.	Merely focuses on throughput performance while ignoring other performances like latency.
	[3]	The distributed algorithm is more efficient than most centralized algorithms.	The caching functionality is not fully related to the computing and networking functionalities.

References

[1] Y. Chen, S. Zhang, S. Xu, and G. Li, "Fundamental trade-offs on green wireless networks," *IEEE Commun. Mag.*, vol. 49, no. 6, pp. 30–37, Jun. 2011.

[2] C. Meixner, C. Develder, M. Tornatore, and B. Mukherjee, "A survey on resiliency techniques in cloud computing infrastructures and applications," *IEEE Commun. Surv. Tuts.*, vol. 18, no. 3, pp. 2244–2281, Feb. 2016.

[3] C. Liang and F. Yu, "Wireless network virtualization: A survey, some research issues and challenges," *IEEE Commun. Surv. Tuts.*, vol. 17, no. 1, pp. 358–380, Aug. 2014.

[4] S. Bu, F. Yu, Y. Cai, and P. Liu, "When the smart grid meets energy efficient communications: Green wireless cellular networks powered by the smart grid," *IEEE Trans. Wireless Commun.*, vol. 11, no. 8, pp. 3014–3024, Aug. 2012.

[5] F. Capozzi, G. Piro, L. Grieco, G. Boggia, and P. Camarda, "Downlink packet scheduling in lte cellular networks: Key design issues and a survey," *IEEE Commun. Surv. Tuts.*, vol. 15, no. 2, pp. 678–700, Jun. 2013.

[6] 3rd Generation Partnership Project (3GPP). (Sep. 2013) Technical specification group services and system aspects; policy and charging control architecture, TS 23.203 V12.2.0. [Online]. Available: http://www.3gpp.org/ftp/specs/html-INFO/23203.htm

[7] Y. Li, H. Xie, Y. Wen, C. Chow, and Z. Zhang, "How much to coordinate? optimizing in-network caching in content-centric networks," *IEEE Trans. Netw. and Service Management*, vol. 12, no. 3, pp. 420–434, Sept. 2015.

[8] A. Ioannou and S. Weber, "A survey of caching policies and forwarding mechanisms in information-centric networking," *IEEE Commun. Surv. Tuts.*, vol. PP, no. 99, pp. 1–41, 2016, IEEE Early Access Articles.

[9] D. Huang, P. Wang, and D. Niyato, "A dynamic offloading algorithm for mobile computing," *IEEE Trans. Wireless Commun.*, vol. 11, no. 6, pp. 1991–1995, Jun. 2012.

[10] O. Munoz, A. Pascual-Iserte, and J. Vidal, "Optimization of radio and computational resources for energy efficiency in latency-constrained application offloading," *IEEE Trans. Vehicular Technology*, vol. 64, no. 10, pp. 4738–4755, Oct. 2015.

[11] W. Zhang, Y. Wen, K. Guan, D. Kilper, H. Luo, and D. Wu, "Energy optimal mobile cloud computing under stochastic wireless channel," *IEEE Trans. Wireless Commun.*, vol. 12, no. 9, pp. 4569–4581, Sept. 2013.

[12] S. Sardellitti, G. Scutari, and S. Barbarossa, "Joint optimization of radio and computational resources for multicell mobile-edge computing," *IEEE Trans. Signal Inf. Process. Over Netw.*, vol. 1, no. 2, pp. 89–103, Jun. 2015.

[13] J. Kwak, Y. Kim, J. Lee, and S. Chong, "Dream: Dynamic resource and task allocation for energy minimization in mobile cloud systems," *IEEE J. Sel. Areas Commun.*, vol. 33, no. 12, pp. 2510–2523, Dec. 2015.

[14] X. Chen, L. Jiao, L. Wenzhong, and F. Xiaoming, "Efficient multi-user computation offloading for mobile-edge cloud computing," *IEEE/ACM Trans. Netw.*, vol. 24, no. 5, pp. 2795–2808, Oct. 2016.

[15] X. Chen, "Decentralized computation offloading game for mobile cloud computing," *IEEE Trans. Parallel and Dist. Systems*, vol. 26, no. 4, pp. 974–983, Apr. 2015.

[16] Y. Cai, F. Yu, and S. Bu, "Dynamic operations of cloud radio access networks (c-ran) for mobile cloud computing systems," *IEEE Trans. Veh. Technol.*, vol. 65, no. 3, pp. 1536–1548, Mar. 2016.

[17] Y. Mao, J. Zhang, and K. Letaief, "Dynamic computation offloading for mobile-edge computing with energy harvesting devices," *IEEE J. Sel. Areas Commun.*, vol. PP, no. 99, pp. 1–16, 2016, IEEE Early Access Articles.

[18] N. Kumar, S. Zeadally, and J. Rodrigues, "Vehicular delay-tolerant networks for smart grid data management using mobile edge computing," *IEEE Commun. Mag.*, vol. 54, no. 10, pp. 60–66, Oct. 2016.

[19] K. Poularakis and L. Tassiulas, "Code, cache and deliver on the move: A novel caching paradigm in hyper-dense small-cell networks," *IEEE Trans. Mobile Computing*, vol. PP, no. 99, pp. 1–14, Jun. 2016, IEEE Early Access Articles.

[20] A. Khreishah, J. Chakareski, and A. Gharaibeh, "Joint caching, routing, and channel assignment for collaborative small-cell cellular networks," *IEEE J. Sel. Areas Commun.*, vol. 34, no. 8, pp. 2275–2284, Aug. 2016.

[21] D. Das and A. Abouzeid, "Co-operative caching in dynamic shared spectrum networks," *IEEE Trans. Wireless Commun.*, vol. 15, no. 7, pp. 5060–5075, July 2016.

[22] D. Liu and C. Yang, "Energy efficiency of downlink networks with caching at base stations," *IEEE J. Sel. Areas Commun.*, vol. 34, no. 4, pp. 907–922, Apr. 2016.

[23] S. Wang, J. Bi, J. Wu, and A. Vasilakos, "CPHR: In-network caching for information-centric networking with partitioning and hash-routing," *IEEE/ACM Trans. Netw.*, vol. 24, no. 5, pp. 2742–2755, Oct. 2016.

[24] X. Yuan, X. Wang, J. Wang, Y. Chu, C. Wang, J. Wang, M. Montpetit, and S. Liu, "Enabling secure and efficient video delivery through encrypted in-network caching," *IEEE J. Sel. Areas Commun.*, vol. 34, no. 8, pp. 2077–2090, Aug. 2016.

[25] P. Si, H. Yue, Y. Zhang, and Y. Fang, "Spectrum management for proactive video caching in information-centric cognitive radio networks," *IEEE J. Sel. Areas Commun.*, vol. 34, no. 8, pp. 2247–2259, Aug. 2016.

[26] D. Malak, M. Al-Shalash, and J. Andrews, "Optimizing content caching to maximize the density of successful receptions in device-to-device networking," *IEEE Trans. Commun.*, vol. 64, no. 10, pp. 4365–4380, Oct. 2016.

[27] L. Zhang, M. Xiao, G. Wu, and S. Li, "Efficient scheduling and power allocation for d2d-assisted wireless caching networks," *IEEE Trans. Commun.*, vol. 64, no. 6, pp. 2438–2452, Jun. 2016.

[28] M. Gregori, J. Gmez-Vilardeb, J. Matamoros, and D. Gndz, "Wireless content caching for small cell and d2d networks," *IEEE J. Sel. Areas Commun.*, vol. 34, no. 5, pp. 1222–1234, May 2016.

[29] A. Liu and V. Lau, "Exploiting base station caching in mimo cellular networks: Opportunistic cooperation for video streaming," *IEEE Trans. Signal Processing*, vol. 63, no. 1, pp. 57–69, Jan. 2015.

[30] K. Poularakis, G. Iosifidis, I. Pefkianakis, L. Tassiulas, and M. May, "Mobile data offloading through caching in residential 802.11 wireless networks," *IEEE Trans. Netw. and Service Management*, vol. 13, no. 1, pp. 71–84, Mar. 2016.

[31] S. Chae and W. Choi, "Caching placement in stochastic wireless caching helper networks: Channel selection diversity via caching," *IEEE Trans. Wireless Commun.*, vol. 15, no. 10, pp. 6626–6637, Oct. 2016.

[32] S. Wang and S. Dey, "Adaptive mobile cloud computing to enable rich mobile multimedia applications," *IEEE Trans. Multimedia*, vol. 15, no. 4, pp. 870–883, June 2013.

[33] C. Zhu, H. Wang, X. Liu, L. Shu, L. Yang, and V. Leung, "A novel sensory data processing framework to integrate sensor networks with mobile cloud," *IEEE Syst. J.*, vol. 10, no. 3, pp. 1125–1136, Sept. 2016.

[34] P. Zhang, Z. Yan, and H. Sun, "A novel architecture based on cloud computing for wireless sensor network," in *Proc. 2nd Int. Conf. Comput. Sci. Electron. Eng.*, 2013.

[35] C. Zhu, V. Leung, L. Yang, and L. Shu, "Collaborative location-based sleep scheduling for wireless sensor networks integrated with mobile cloud

computing," *IEEE Trans. Computers*, vol. 64, no. 7, pp. 1844–1856, July 2015.

[36] W. Zhang, Y. Wen, and D. Wu, "Collaborative task execution in mobile cloud computing under a stochastic wireless channel," *IEEE Trans. Wireless Commun.*, vol. 14, no. 1, pp. 81–93, Jan. 2015.

[37] C. Li and H. Dai, "Efficient in-network computing with noisy wireless channels," *IEEE Trans. Mobile Computing*, vol. 12, no. 11, pp. 2167–2177, Nov. 2013.

[38] C. You, K. Huang, and H. Chae, "Energy efficient mobile cloud computing powered by wireless energy transfer," *IEEE J. Sel. Areas Commun.*, vol. 34, no. 5, pp. 1757–1771, May 2016.

[39] Z. Yin, F. R. Yu, S. Bu, and Z. Han, "Joint cloud and wireless networks operations in mobile cloud computing environments with telecom operator cloud," *IEEE Trans. Wireless Commun.*, vol. 14, no. 7, pp. 4020–4033, July 2015.

[40] M. Kangas, S. Glisic, Y. Fang, and P. Li, "Resource harvesting in cognitive wireless computing networks with mobile clouds and virtualized distributed data centers: performance limits," *IEEE Trans. Cognitive Commun. and Netw.*, vol. 1, no. 3, pp. 318–334, Sept. 2015.

[41] X. Liu, Y. Li, and H. Chen, "Wireless resource scheduling based on backoff for multiuser multiservice mobile cloud computing," *IEEE Trans. Veh. Technol.*, vol. 65, no. 11, pp. 9247–9259, Nov. 2016.

[42] R. Chiang, S. Rajasekaran, N. Zhang, and H. Huang, "Swiper: Exploiting virtual machine vulnerability in third-party clouds with competition for i/o resources," *IEEE Trans. Parallel and Distributed Systems*, vol. 26, no. 6, pp. 1732–1742, Jun. 2015.

[43] R. Deng, R. Lu, C. Lai, T. Luan, and H. Liang, "Optimal workload allocation in fog-cloud computing toward balanced delay and power consumption," *IEEE Internet of Things Journal*, vol. 3, no. 6, pp. 1171–1181, Dec. 2016.

[44] L. Pu, X. Chen, J. Xu, and X. Fu, "D2d fogging: An energy-efficient and incentive-aware task offloading framework via network-assisted d2d collaboration," *IEEE J. Sel. Areas Commun.*, vol. 34, no. 12, pp. 3887–3901, Dec. 2016.

[45] X. Hou, Y. Li, M. Chen, D. Wu, D. Jin, and S. Chen, "Vehicular fog computing: A viewpoint of vehicles as the infrastructures," *IEEE Trans. Veh. Technol.*, vol. 65, no. 6, pp. 3860–3873, Jun. 2016.

[46] F. Jalali, K. Hinton, R. Ayre, T. Alpcan, and R. Tucker, "Fog computing may help to save energy in cloud computing," *IEEE J. Sel. Areas Commun.*, vol. 34, no. 5, pp. 1728–1739, May 2016.

[47] R. Gargees, B. Morago, R. Pelapur, D. Chemodanov, P. Calyam, Z. Oraibi, Y. Duan, G. Seetharaman, and K. Palaniappan, "Incident supporting visual cloud computing utilizing software-defined networking," *IEEE Trans. on Circuits and Systems for Video Technology*, pp. 1–15, 2016, IEEE Early Access Articles.

[48] D. Zeng, L. Gu, S. Guo, Z. Cheng, and S. Yu, "Joint optimization of task scheduling and image placement in fog computing supported software-defined embedded system," *IEEE Trans. Computers*, vol. 65, no. 12, pp. 3702–3712, Dec. 2016.

[49] S. Cirani, G. Ferrari, N. Iotti, and M. Picone, "The IoT hub: a fog node for seamless management of heterogeneous connected smart objects," in *Proc. 12th Annual IEEE International Conference on Sensing, Communication, and Networking - Workshops* (SECON Workshops), Seattle, WA, USA, Jun. 2015, pp. 1–6.

[50] G. Lee, W. Saad, and M. Bennis, "An online secretary framework for fog network formation with minimal latency," *arXiv preprint arXiv:1702.05569 (2017)*, 2017.

[51] Y. Wang, M. Sheng, X. Wang, L. Wang, and J. Li, "Mobile-edge computing: Partial computation offloading using dynamic voltage scaling," *IEEE Trans. Commun.*, vol. 64, no. 10, pp. 4268–4282, Oct. 2016.

[52] C. Wang, F. Yu, C. Liang, Q. Chen, and L. Tang, "Joint computation offloading and interference management in wireless cellular networks with mobile edge computing," *IEEE Trans. Veh. Technol.*, vol. 66, no. 8, pp. 7432–7445, Mar. 2017.

[53] C. Wang, F. R. Yu, Q. Chen, and L. Tang, "Joint computation and radio resource management for cellular networks with mobile edge computing," in *Proc. IEEE ICC17*, Paris, France, May 2017.

[54] N. Master, A. Dua, D. Tsamis, J. Singh, and N. Bambos, "Adaptive prefetching in wireless computing," *IEEE Trans. Wireless Commun.*, vol. 15, no. 5, pp. 3296–3310, May 2016.

[55] Y. Jin, Y. Wen, and C. Westphal, "Optimal transcoding and caching for adaptive streaming in media cloud: an analytical approach," *IEEE Trans. Circuits and Systems for Video Technol.*, vol. 25, no. 12, pp. 1914–1925, Dec. 2015.

[56] G. Gao, W. Zhang, Y. Wen, Z. Wang, and W. Zhu, "Towards cost efficient video transcoding in media cloud: Insights learned from user viewing

patterns," *IEEE Trans. Multimedia,* vol. 17, no. 8, pp. 1286–1296, Aug. 2015.

[57] A. Khanafer, M. Kodialam, and K. Puttaswamy, "To rent or to buy in the presence of statistical information: The constrained ski-rental problem," *IEEE/ACM Trans. Netw.,* vol. 23, no. 4, pp. 1067–1077, Aug. 2015.

[58] J. Liu, Y. Zhou, and D. Zhang, "Transim: A simulation framework for cache-enabled transparent computing systems," *IEEE Trans. Computers,* vol. 65, no. 10, pp. 3171–3183, Oct. 2016.

[59] E. Bastug, M. Bennis, E. Zeydan, M. Kader, I. Karatepe, A. Er, and M. Debbah, "Big data meets telcos: A proactive caching perspective," *J. Commun. Netw.,* vol. 17, no. 6, pp. 549–557, Dec. 2015.

[60] X. Dai, X. Wang, and N. Liu, "Optimal scheduling of data-intensive applications in cloud-based video distribution services," *IEEE Trans. Circuits and Systems for Video Tech.,* vol. 27, no. 1, pp. 73–83, Jan. 2017.

[61] J. Mukherjee, D. Krishnamurthy, and J. Rolia, "Resource contention detection in virtualized environments," *IEEE Trans. Netw. and Service Management,* vol. 12, no. 2, pp. 217–231, Jun. 2015.

[62] M. Kiskani and H. Sadjadpour, "Throughput analysis of decentralized coded content caching in cellular networks," *IEEE Trans. Wireless Commun.,* vol. 16, no. 1, pp. 663–672, Jan. 2017.

[63] C. Wang, C. Liang, F. Yu, Q. Chen, and L. Tang, "Computation offloading and resource allocation in wireless cellular networks with mobile edge computing," *IEEE Trans. Wireless Commun.,* vol. 16, no. 8, pp. 4924–4938, May 2017.

[64] C. Wang, C. Liang, F. R. Yu, Q. Chen, and L. Tang, "Joint computation offloading, resource allocation and content caching in cellular networks with mobile edge computing," in *Proc. IEEE ICC17,* Paris, France, May 2017.

Chapter 3

Edge Caching with Wireless Software-Defined Networking

Software-defined networking (SDN) and in-network caching are promising technologies in next generation wireless networks. In this chapter, we propose to enhance quality of experience (QoE)-aware wireless edge caching with bandwidth provisioning in software-defined wireless networks (SDWNs). Specifically, we design a novel mechanism to jointly provide proactive caching, bandwidth provisioning and adaptive video streaming. The caches are requested to retrieve data in advance dynamically according to the behaviors of users, the current traffic and the resource status. Then, we formulate a novel optimization problem regarding the QoE-aware bandwidth provisioning in SDWNs with jointly considering in-network caching strategy. The caching problem is decoupled from the bandwidth provisioning problem by deploying the dual-decomposition method. Additionally, we relax the binary variables to real numbers so that those two problems are formulated as a linear problem and a convex problem, respectively, which can be solved efficiently. Simulation results are presented to show that latency is decreased and the utilization of caches is improved in the proposed scheme.

3.1 Wireless SDN and edge caching

According to [1], global mobile data traffic will increase nearly eightfold from 2015 to 2020 and 75 percent of it will be video. Moreover, another prominent feature of next generation wireless networks would be the full support of *software-defined networking* (SDN)[2] in wireless networks[3]. These trends enforce the optimization of wireless networks to jointly consider the improvement of QoE for video services and the integration of SDN. New networking technologies in wireless networks, such as *heterogeneous networks* (HetNets) [4, 5], *software-defined wireless networks* (SDWNs) [6] and *wireless edge caching* (WEC) [7, 8] that are arising to solve these changes have been proposed and studied recently.

Generally speaking, SDWNs can enable the reduction of complexity and cost of networks, equip programmability into wireless networks, accelerate evolution of networks, and even further catalyze fundamental changes in the mobile ecosystem[3]. The success of the SDWN will depend critically on our ability to jointly provision the backhaul and radio access networks (RANs) [9].

WEC, as an extension of *in-network caching* that can efficiently reduce duplicate content transmission [10], has shown that access delays, traffic loads, and network costs can be potentially reduced by caching contents in wireless networks [11]. To successfully combine caching and wireless networks, significant work (e.g., [12, 13]) has been done concerning utilizing and placing contents in the caches of base stations (BSs).

3.1.1 Motivations and contributions

As the QoE of streaming video services mainly includes video resolutions, buffering delays and stalling events [14, 15], the SDWN (e.g., provision QoS [16]) and wireless edge caching (e.g., reduce delay [17]) appear as promising candidates to enhance QoE. However, to the best of our knowledge, QoE-aware joint optimization of the network and cache resources in SDWNs has been largely ignored in the existing research. Unfortunately, the combination of those issues is not straightforward, as several challenges are induced by this joint optimization as follows. First, bandwidth provisioning in SDWNs should be content-aware, which means it should assign network resources to users based on caching status and improvement of the QoE. Second, to enhance the hitting ratio of caches (utilization), caching strategies should be proactive according to the current traffic and resource status, behaviors of users, as well as the requirements of the QoE. Third, since video SDN flows from service providers usually have minimum requirements, the overall QoE performance of the network needs to be guaranteed.

Thus, to address those issues in this chapter, we propose to jointly optimize QoE of video streaming, bandwidth provisioning and caching strategies in

SDWNs with limited network resources and QoE requirements. The distinctive technical features of this chapter are as follows:

- To decrease content delivery latency and improve utilization of the network resources and caches, we design a novel mechanism to jointly provide proactive caching, bandwidth provisioning and adaptive video streaming. BSs are requested to retrieve data in advance dynamically according to the behaviors of users, the current traffic and the resource status.

- To cope with the limited resources and the quality of service requirements, we formulate a novel optimization problem regarding QoE-aware bandwidth provisioning in SDWNs with jointly considering an in-network caching strategy.

- The caching problem is decoupled from the bandwidth provisioning problem by deploying the dual-decomposition method. Additionally, we relax the binary variables to real numbers so that those two problems are formulated as a linear problem and a convex problem, respectively, which can be solved efficiently.

- Algorithms are proposed to achieve the sub-optimum solution by solving the relaxed problem and utilizing a rounding up method to recover relaxed variables to binary.

3.1.2 Literature review

Bandwidth provisioning (flow control) of SDWNs is studied in [16, 18] through traffic engineering. The authors of [18] propose a multi-path traffic engineering formulation for downlink transmission considering both backhaul and radio access constraints. Moreover, the link buffer status is used as feedback to assist the adjustment of flow allocation. Based on [18], the authors of [16] extend flow control of SDWNs to real-time video traffic specifically. This research proposes an online method to estimate the effective rate of video flow dynamically. Minimum flow rate maximization of SDWNs is investigated in [9] with jointly considering flow control and physical layer interference management problems using a weighted-minimum mean square error algorithm.

In the line of caching at mobile networks, the authors of [19] formulate a delay minimization problem by optimally caching contents at the SBS. This research first clusters users with similar content and then deploys a reinforcement learning algorithm to optimize its caching strategy accordingly. [19] proposes a scheme that BSs opportunistically employ cooperative multiple-input multiple-output (Coop-MIMO) transmission by caching a portion of files so that a MIMO cooperation gain can be achieved without payload backhaul, while [13, 1] propose to utilize caching for releasing part of the fronthaul in C-RAN. Unfortunately, the research does not take real-time mobile network and traffic status into account. In [21], the proposed cache allocation policy considers both the backhaul consumption of small BSs (SBSs) and local storage constraints. The authors of [22] introduce in-network caching into SDWNs to reduce the latency of the backhaul, but cache strategies are ignored in this study. [23] introduces dynamic caching into BS selection from the aspect of energy saving. In [24], dynamic caching is proposed to consider the mobility of users, but it focuses more on the cache resource and relationship between nodes. Network resources, such as the backhaul capacity and spectrum, are not discussed in this research.

Video quality adaptation with radio resource allocation in mobile networks (especially long term evolution (LTE)) is studied in a number of studies. The authors of [25] provide a comprehensive survey on quality of experience of HTTP adaptive streaming. Joint optimization of video streaming and in-network caching in HetNets can be traced back to [17]. This research suggests that SBSs form a wireless distributed caching network that can efficiently transmit video files to users. Meanwhile, Ahlehagh and Dey [26] take the backhaul and the radio resource into account to realize video-aware caching strategies in RANs for the assurance of maximizing the number of concurrent video sessions. The authors of [27] move the attention of in-network video caching to core networks, instead of RANs, of LTE. To utilize new technologies in next generation networks, [28] studies the quality of video in next generation networks. [14] conducts comprehensive research that investigates opportunities and challenges of combining the advantages of adaptive bit rate and RAN caching to increase the video capacity and QoE of wireless networks. Another study combining caching and video service is [29], where collaborative caching is studied with jointly considering scalable video coding. However, RAN and overall video quality requirements are not considered.

3.2 System model and problem formulation

In this section, we present the system model of the HetNet, dynamic caching, service QoE and related assumptions. The notations that will be used in the rest of this chapter are summarized in Table 3.1.

3.2.1 Network Model

3.2.1.1 Wireless communication model

In this chapter, we consider the downlink transmission case in a software-defined HetNet comprosed of a set $\mathcal{J}$ of cache-enabled BSs, such as macro BSs (MBSs) and small BSs (SBSs). The area covered by SBSs overlaps that covered by MBSs. A central SDN controller is deployed to control the network including caching strategies and bandwidth provisioning. BSs connect to the CN through wired backhaul links with fixed capacities (e g , bps). A content

Table 3.1: Notations

Notation	definition
$\mathcal{J}$	the set of mobile BS nodes (e.g., MBSs, SBSs)
$J = \mid \mathcal{J} \mid$	number of BSs
j,l	BS, $j = 0$ means content cloud library
$a_{lj} \in \{0,1\}$	link status between BS j and BS l
R_{lj}^{max} (bps)	backhaul link capacity between BS j and BS l
$\mathcal{I}_j, \mathcal{I}$	the set of users
$\mathcal{I}_j^{\pi}$	the predicted (potential) set of users served by BS j
i	wireless user
g_{ij}	large scale channel gain between i and BS j
σ_0	noise power spectrum density
p_j (Watts / Hz)	normalized transmission power of BS j
w_{ij} (Hz)	assigned spectrum bandwidth of user i from j
W_j (Hz)	total spectrum bandwidth of j
γ_{ij}	received SINR of user i served by BS j
Γ_{ij}	spectrum efficiency of user u served by BS j
C	the total number of contents
C_j	the storage capacity of BS j
$h_{ij} \in \{0,1\}$	hitting event indicator between user i and BS j
$c_{ij} \in \{0,1\}$	dynamic caching indicator
$\tilde{c}_{ij} \in [0,1]$	relaxed dynamic caching indicator
δ_0 (seconds)	evaluation time (scheduling cycle time)
π_{ij}	predicted probability of user i served by BS j
Q	number of total video resolution levels
$q \in \{1,...,6\}$	video resolution level indicator
s_q	the MV-MOS of the resolution q
v_q	required transmission data rate of the resolution q
$x_{qi} \in \{0,1\}$	resolution indicator of user i
$\tilde{x}_{qi} \in [0,1]$	relaxed resolution indicator of user i
d_j	the average backhaul transmission delay of BS j
b_0 (seconds)	minimum buffer length to play video
b_{qi} (seconds)	current buffer length of resolution q at user u
S_0 (seconds)	the MV-MOS requirements from service providers
U	*Total utilities*
ι	consistence
r_{ij}^c (bps)	required data transmission rate of caching
r_{lj}^f (bps)	provisioned bandwidth for backhaul links
r_{ij}^a (bps)	provisioned bandwidth for air interfaces
λ, μ, ν	dual variables
G, L_j	dual functions
Ω_{ij}	caching margin gain
N_j	number of cachable contents

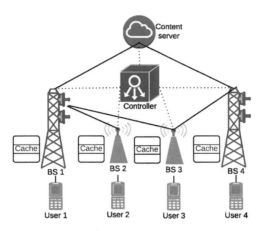

Figure 3.1: Network architecture of the cache-enabled SDWN.

server is physically located at a core router (content delivery networks) in the CN or the source server. Moreover, as shown in Figure. 3.1, some BSs (e.g., SBSs) connect to the CN through MBSs. Without loss of generality, each BS may have multiple links to be connected in the networks (e.g., SBS 3 in Fig. 3.1). We use link indicator a_{lj} to denote the link status between BS l and BS j. $a_{lj} = 1$ means BS l is the next hop of BS j with a fixed link capacity R_{lj}^{max} (bps); otherwise $a_{lj} = 0$. Since it is the downlink case, $a_{lj} = 1$ implies $a_{jl} = 0$. Specially, we use $j = 0$ to indicate the CN. Let $r_{lj}^f \in \mathbb{R}^+$ denote the provisioning bandwidth of BS j for its next hop BS l through wired backhaul links. f in the superscript is used to denote forwarding. The following constraints (backhaul limitation) holds

$$r_{lj}^f \le R_{lj}^{max}, \forall j, l \in \mathcal{J}. \tag{3.1}$$

Let $\mathcal{I}_j$ denote the set of users served by BS j and each user i requests a video flow with certain data rate requirements. The set $\mathcal{I} := \mathcal{I}_1 \cup ... \cup \mathcal{I}_j$ of users means the total users set. To enforce single association, we let $\mathcal{I}_1 \cap \mathcal{I}_j = \varnothing$ and any user is served by the BS that provides the best spectrum efficiency (SE). Association schemes are well-studied in heterogeneous wireless networks [30, 31]. In this chapter, to simplify our analysis, we do not consider any advanced interference management and power allocation schemes. We assume that the spectrum used by different users within one BS is orthogonal, which means there is no intra-cell interference between users associated with the same BS. This is similar to a system of frequency-division multiple access, such as LTE. The spectrum reuse factor for all cells is one (overlaid), which means inter-cell downlink interference is considered. A fixed equal power allocation mechanism is used, where the normalized transmit power on BS j is p_j Watts/Hz regardless of positions and the allocated spectra of users.

If $i \in \mathcal{I}_j$, the signal-to-interference plus noise (SINR) γ_{ij} can be determined by

$$\gamma_{ij} = \frac{g_{ij} p_j}{\sigma_0 + \sum_{l, l \neq j} g_{il} p_l} \tag{3.2}$$

where g_{ij} is the channel gain between user i and BS j including large-scale path loss and shadowing, σ_0 is the power spectrum density of additive white Gaussian noise and $\sum_{l, l \neq j} g_{il} p_l$ is the aggregated received interference. As small-scale fading varies much faster than caching and bandwidth provisioning, small-scale fading is not considered when evaluating the SINR. Therefore, the SINR calculated by (3.2) can be considered an average SINR. Moreover, in this chapter, as we mainly consider the benefits from dynamic caching and flow-level scheduling, radio resource allocation is not considered, which leads to a formulation where small-scale fading is ignored. However, small-scale fading can have effects on the SINR if the caching process is comparably fast. Thus, if small-scale fading is considered, the network link capacity (e.g., γ_{ij}) can be modeled as random variables, which leads to a stochastic optimization problem. The scheme in [32] can provide a suitable solution to this problem.

By using the Shannon bound to get the spectrum efficiency (SE) $\Gamma_{ij} = \log(1 + \gamma_{ij})$, the provisioning (achievable) radio access data rate $r_{ij}^a \in \mathbb{R}^+$ for user i with BS j can be calculated as:

$$r_{ij}^a - w_{ij} \log(1 + \gamma_{ij}) \tag{3.3}$$

where w_{ij} (Hz) is the available spectrum bandwidth of BS j allocated to user i. a in the superscript is used to denote the air interface. Since the spectrum bandwidth of each BS j is limited by W_j, the following constraints must hold:

$$\sum_{i \in \mathcal{I}_j} \frac{r_{ij}^a}{\Gamma_{ij}} \leq W_j, \forall j \in \mathcal{J}. \tag{3.4}$$

3.2.1.2 Proactive wireless edge caching model

To improve caching performance, BSs can proactive cache contents that may be requested by users. The probability that a cached content in BS j will be used by user i is assumed to be π_{ij}. In practice, π_{ij} depends on the user mobility pattern [33] (current location, moving velocity and direction, and unpredictable factors) and the popularity of the content [24]. Research about the behavior of users has attracted great interest from both academia and industry [34]. Necessary information for processing a prediction can come from real-time data (e.g., UE and RANs) and historical data (e.g., users database). Fortunately, the improvement of machine learning and big data technologies can help the management and control of mobile networks [35, 36]. By using those advanced techniques, the network is able to have a deeper view of the traffic and the coverage from the huge volume of historical data, which may help the network to enhance the accuracy of the prediction. It should be noted that the calculation of this probability is beyond the research of this chapter and we assume that π_{ij} is known in advanced by the SDWN controller (before each scheduling cycle starts). To evaluate the effect of π_{ij}, we test our performance with two different π_{ij} setups in the simulation. Let us use $\mathcal{I}_j^\pi$ to denote the potential set of users that may be served by BS j, which can be formally defined as $\mathcal{I}_j^\pi := \{i \mid \pi_{ij} \neq 0, i \in \mathcal{I}\}$.

In this proposed scheme, we assume the total amount of content C is stored at the content server (cloud centers or the Internet source) and each content has the normalized size of 1. This assumption is reasonable because we can slice the video content into chunks of the same length. Our BSs are assumed to be cache-enabled, so there is a physical cache at each BS n with capacity C_j. As any BS only has limited storage capability, C_j is much smaller than the cloud center ($C_j \ll C, \forall j$) and the cache placement decision is made by the SDN controller. We assume that the SDN controller knows all the information about content retrieving requests from users and the cache placement status of all BSs. Let us use a binary parameter h_{ij} to denote a hitting event between user i and BS j. $h_{ij} = 1$ means the content requested by user i can be found at BS j and $h_{ij} = 0$ means the opposite. According to the control signaling by the SDN controller, contents are pulled actively by BSs through wired backhaul links. An example is shown in Section 3.3.5.

Denote $c_{ij} \in \{0, 1\}$ to be the binary decision variable used to control whether the content i (potentially requested by user i) is going to be placed at BS j or not. The bandwidth required to cache the content requested by i is pre-defined as r_i^c (bps) that is fixed. Hence, the provisioned bandwidth for caching in BS j is $\sum_{i \in \mathcal{I}_j^\pi} c_{ij} r_i^c$. Note duplicated caching is avoided as the SDN controller knows the cached contents of every BS. In other words, if the content has been cached in BS j, it is unnecessary to cache again, namely, $c_{ij} = 0$ if $h_{ij} = 1$. Since each BS j has limited cache space, constraints should hold to ensure the caching strategy is limited in the empty space of the cache of each access BS j.

$$\sum_{i \in \mathcal{I}_j^\pi} c_{ij} \leq C_j, \forall j \in \mathcal{J}. \tag{3.5}$$

3.2.1.3   Video QoE model

In this chapter, to specify network performance for video services, we use a novel video experience evaluation method called MV-MOS proposed in [37] to model network utility. MV-MOS is a more advanced measurement of the video quality based on the well-known video mean opinion score (vMOS), and it can adapt to the evolution of video resolutions. We assume the q-th resolution of any video requires a data rate v_q (bps) and gains a mobile video mean-opinion-score (MV-MOS) s_q (from 1 to 6). For example, [37] points out that a 4k video average requires a 25Mbps data rate and can be quantified to a MV-MOS 4.9 out of 5. Practically, v_q depends on video coding schemes and the video content, which vary with time. Nevertheless, since the purpose of our chapter is to maximize the wireless network performance dynamically instead of video services, we assume the required data rate is a fixed value v_q over research time for all streams.

We define x_{iq} as the resolution indicator of user i and Q is the highest resolution level. Specifically, if the q-th resolution of the video is selected by user i, $x_{iq} = 1$; otherwise, $x_{iq} = 0$. Thus, for any user i, the experienced MV-MOS is $\sum_q x_{iq} s_q$ and the required data rate is $\sum_q x_{iq} v_q$. Since users can select only one level of resolution at the same time, the following constraints should hold:

$$\sum_{q=1}^{Q} x_{iq} = 1. \forall i \in \mathcal{I} \tag{3.6}$$

Moreover, to guarantee the video service QoE of the overall network, we should guarantee the overall average MV-MOS higher than an acceptable value S_0, namely,

$$\frac{1}{I} \sum_{i \in \mathcal{I}} \sum_{q=1}^{Q} s_q x_{iq} \geq S_0. \tag{3.7}$$

Usually, the S_0 is provided by service providers or network operators to control the overall performance of a certain area.

In [14] and [15], the authors point out that stalling (video interruption) degrades users experience more severely than initial latency because it disrupts the smoothness of streaming videos. This happens when the buffer b_{iq} (bits) is exhausted, lower than a threshold b_0. In order to avoid this "annoyance" factor, in this chapter, we request the wireless network to maintain the buffer higher than a threshold b_0 (bits) at the user device buffer, so that the user can get a smooth streaming video experience. This proactive buffer management that feed back the buffer status of users to the network has been proposed in existing research [16, 38, 39]. To maintain b_0 of video, for any user i, it should follow the constraints:

$$\delta \sum_{j \in \mathcal{J}} r_{ij}^a + \sum_{q=1}^{Q} x_{iq} b_{iq} - \delta \sum_{q=1}^{Q} x_{iq} v_q \geq b_0, \forall i \in \mathcal{I} \tag{3.8}$$

where δ is a predefined time window unit in seconds. Every δ seconds, the system adaptively selects the resolution of the demanded video based on buffer status b_{iq} and available data rate $\sum_{j \in \mathcal{J}} r_{ij}^a$. (3.8) means that the amount of downloaded data and buffered data must support playback average δ (s) and maintain at least b_0 buffer.

3.2.2 Problem formulation

In this subsection, an optimization problem maximizing the gains from wireless edge caching constrained by physical resource and service QoE is proposed.

The purpose of this study is to improve overall wireless edge caching performance by considering caching strategies, network status and QoE provisioning, so we have to choose an appropriate network utility function U. First, the caching utilization should be represented by different metrics (e.g., alleviated backhaul, delay reduction, and reduced network costs). Second, as this study is QoE-aware where delay is one of the most important metrics, we chose delay reduction* d_j (seconds) as the gain of caching at the wireless edge [17]. Thus, as shown next, we define a logarithm-based objective function as the utility function, which is equivalent to the proportional fairness scheme in a long term view [40]:

$$U(\mathbf{c}) = \sum_{j \in \mathcal{J}} \log \left(\sum_{i \in \mathcal{I}_j^\pi} \pi_{ij} \bar{h}_{ij} c_{ij} d_j + \iota \right) \tag{3.9}$$

where $\bar{h}_{ij} = 1 - h_{ij}$ means the absence of the content requested by user i and $\iota \geq 1$ is consistences to guarantee $U(\mathbf{c})$ does not fall into negative infinity.† The product of $\bar{h}_{ij}$ and c_{ij} enforces c_{ij} to be different from h_{ij} if $h_{ij} = 1$, which is equivalent to the case that duplicated caching is avoided. The probability π_{ij} can be interpreted as a weighted factor that represents the success of caching. Therefore, $\pi_{ij} \bar{h}_{ij} d_j$ can be considered the expected reduced backhaul latency, if $c_{ij} = 1$. To lighten the notations, we let $H_{ij} = \pi_{ij} \bar{h}_{ij} d_j$. We select a logarithm-based objective function due to the following features that are perceived [40]:

■ *Component-wise monotonic increase* means that larger caching gain yields larger utility;

■ *Convexity* can guarantee the convergence and efficiency of the algorithm;

■ *Fairness-awareness* gives a desired balance between the overall network and the single user.

To utilize the physical resource efficiently, the total provisioned bandwidth of flows from BS j should be less than the total bandwidth of flows coming into the corresponding BS. Otherwise, allocated spectrum or backhaul links will be left unused. The flow conservation constraint for BS j can be written as

$$\underbrace{\overbrace{\sum_{l \in \mathcal{J}} a_{lj} r_{lj}^f}^{\text{forwarding}} + \overbrace{\sum_{i \in \mathcal{I}_j} \bar{h}_{ij} r_{ij}^a}^{\text{radio}} + \overbrace{\sum_{i \in \mathcal{I}_j^\pi} c_{ij} r_i^c}^{\text{caching}}}_{\text{out flows}} \leq \underbrace{\sum_{l \in \mathcal{J} \cup 0} a_{jl} r_{jl}^f}_{\text{in flows}}, \quad \forall j \in \mathcal{J}. \tag{3.10}$$

---

*We use the average backhaul downlink latency as the measurement value.

†When either $\mathcal{I}_j^\pi = \varnothing$ or $\pi_{ij} \bar{h}_{ij} = 0, \forall i \in \mathcal{I}_j^\pi$, the objective function may result in an infeasible problem.

The first term left on the (3.10) is in reserved bandwidth for flows forwarded to the next hop BSs. The second term is the summation of all provisioned bandwidth for bearing users' radio access flows. Obviously, if $\bar{h}_{ij} = 0$ that means we can find existing data (cached in previous caching cycles) on BS j for user i so that backhaul consumption is avoided. The third term is all caching flows. Obviously, the provisioned (reserved) bandwidth of these three kinds of flows cannot be larger than the reserved backhaul bandwidth for this BS. If $\bar{h}_{ij} = 0$, it is reserved for two kinds of flows (caching and forwarding). The left term of (3.10) can be a dynamic backhaul link capacity of BS j.

Thus, given the objective function, and constraints, we can define the joint flow bandwidth provisioning and wireless edge cache placement problem **P0** as follows:

$$\textbf{P0}: \max_{\substack{\mathbf{r}^a, \mathbf{r}^f \in \mathbb{R}+ \\ \mathbf{c}, \mathbf{x} \in \{0,1\}}} \quad U \tag{3.11a}$$

$$s.t. \quad (3.1), (3.4), (3.5), (3.6), (3.7), (3.8), (3.10) \tag{3.11b}$$

Unfortunately, problem (3.11), however, is difficult to solve and implement based on the following observations:

- The mix integer variables result a mix-integer non-linear problem (MINLP) that generally is NP-hard and intractable in scheduling[41];

- The complexity of solving (3.11) by using greedy or genetic methods will increase significantly with the increase of the number of users and (or) BSs. In future cellular networks, the density and number of small cells will rise significantly so that the size of variables will become very large;

- Cache indicators and flow bandwidth are decided by different layers and perform in different time scales.

3.3   Bandwidth provisioning and edge caching

In this section, an algorithm is proposed to solve the problem **P0** in (3.11). As those integer variables block us from finding a traceable algorithm, we first relax binary variables c_{ij} and x_{iq} to real numbers variables $\tilde{c}_{ij}$ and $\tilde{x}_{iq}$ bounded by $[0, 1]$ so that the domain of variables is a convex set. These relaxations are a common way to deal with binary variables in wireless networks[29, 31, 40, 42]. From a long term view, we can interpret this kind of relaxation as a partial allocation or average time sharing portion. For example, a partial caching indicator $\tilde{c}_{ij}$ or resolution $\tilde{x}_{iq}$ means the portion of time when BS j indicates a cache to user i or when user i selects the resolution level q, respectively. Moreover, by solving the relaxed binary variables, the upper bound of the proposed algorithm can be achieved. We evaluate the gap between this upper bound and the final solution in our simulation.

After relaxing binary variables, it is clear that the feasible set of the revised problem (3.11) is a convex set because all constraints are linear constraints and the variables domain is a convex set. Since the log-based objective function is a strict concave function regarding $\tilde{c}_{ij}$, the problem (3.11) is transferred to a convex problem **P̃0** that can be solved effectively without effort by using general solvers. In the remaining subsections, we will present an algorithm based on dual-decomposition to solve the relaxed problem **P̃0** and recover relaxed variables back to binaries.

3.3.1 Proposed caching decoupling via dual decomposition

This study aims to adaptively replace content in caches of BSs while optimizing bandwidth provisioning and video resolution selection. As observed from the problem **P0**, we can use the spare backhaul bandwidth to cache content to BSs before users are actually served (or potentially continued served) by those BSs (or actually request the content) and select a best video resolution for each user based on an allocated achievable data rate. Moreover, caches and bandwidth actually belong to different layers of the network. Usually, flow bandwidth is restricted by physical resources (provisioned by the media access control layer), and x_{iq} depends on the achievable data rate. Joint optimizations of video rate and flow bandwidth are studied in cross-layer design schemes [43]. Alternative, an in-network cache is placed at the network layer or even higher layer. In addition, scheduling of caching can be in a longer time period compared to flow scheduling. Fortunately, constraints coupling cache and bandwidth are only (3.10).

Thus, those features of **P0** motivate us to adopt a dual-decomposition method so that the caching problem can be separated from bandwidth provisioning and resolution selection. We form the partial Lagrangian function for the problem **P̃0** by introducing the dual variables (backhaul prices) $\{\lambda_j\}$ for constraints (3.10). Then, the partial Lagrangian function can be shown as:

$$
\begin{aligned}
G(\lambda) = \sum_{j \in \mathcal{J}} \log \left(\sum_{i \in \mathcal{I}_j^\pi} H_{ij} \tilde{c}_{ij} + \iota \right) - \sum_{j \in \mathcal{J}} \lambda_j \sum_{i \in \mathcal{I}_j^\pi} \tilde{c}_{ij} r_i^c \\
- \sum_{j \in \mathcal{J}} \lambda_j \left(\sum_{l \in \mathcal{J}} a_{lj} r_{lj}^f + \sum_{i \in \mathcal{I}_j} \bar{h}_{ij} r_{ij}^a - \sum_{l \in \mathcal{J} \cup 0} a_{jl} r_{jl}^f \right)
\end{aligned}
\tag{3.12}
$$

The dual problem (DP) is thus:

$$
\mathbf{DP} : \min_{\lambda \in \mathbb{R}^+} \quad G(\lambda) = f_c(\lambda) + f_{r,x}(\lambda)
\tag{3.13}
$$

where

$$
f_c(\lambda) = \underset{c \in [0,1]}{\operatorname{argmax}} \left\{
\begin{array}{l}
\displaystyle \sum_{j \in \mathcal{J}} \log \left(\sum_{i \in \mathcal{I}_j^\pi} H_{ij} \tilde{c}_{ij} + \iota \right) \\[4mm]
\displaystyle - \sum_{j \in \mathcal{J}} \lambda_j \sum_{i \in \mathcal{I}_j^\pi} \tilde{c}_{ij} r_i^c, \\[4mm]
\text{s.t.} \quad (3.5)
\end{array}
\right\} \tag{3.14}
$$

and

$$
f_{r,x}(\lambda) = \underset{\substack{r^a, r^f \in \mathbb{R}^+ \\ x \in [0,1]}}{\operatorname{argmax}} \left\{
\begin{array}{l}
\displaystyle \sum_{l \in \mathcal{J} \cup 0} \sum_{j \in \mathcal{J}} \lambda_j a_{jl} r_{jl}^f - \\[4mm]
\displaystyle \sum_{j \in \mathcal{J}} \lambda_j \left(\sum_{l \in \mathcal{J}} a_{lj} r_{lj}^f + \sum_{i \in \mathcal{I}_j} \bar{h}_{ij} r_{ij}^a \right), \\[4mm]
\text{s.t.} \quad (3.1), (3.4), (3.6), (3.7), (3.8)
\end{array}
\right\} \tag{3.15}
$$

Let us introduce a parameter z_n called extra bandwidth for each BS j shown as:

$$
z_j = \sum_{l \in \mathcal{J} \cup 0} a_{jl} r_{jl}^f - \sum_{\iota \subset \mathcal{I}_j^\pi} \tilde{c}_{ij} r_i^c - \sum_{l \subset \mathcal{J}} a_{lj} r_{lj}^f - \sum_{i \subset \mathcal{I}_j} \bar{h}_{ij} r_{ij}^a. \tag{3.16}
$$

According to dual decomposition[44], a subgradient of $G(\lambda)$ is $z = [z_1, ..., z_j]^T$, and we thus can update λ based on:

$$
\lambda^{[k+1]} = \lambda^{[k]} - \alpha_\lambda^{[k]} z^{[k]}, \tag{3.17}
$$

where $\alpha_\lambda^{[k]}$ is the step length at the iteration step $[k]$.

3.3.2 Upper bound approach to solving (3.14)

In this section, a low complexity algorithm is proposed to solve the problem (3.14) so that the upper bound can be achieved. First, to make our expression more compact, we define $y_{ij} = \lambda_j r_i^c$, which can be interpreted as the backhaul bandwidth cost of caching. Observe that $f_c(\lambda)$ can be further decoupled to each BS j, thus we focus on solving (3.14) for one BS. Let $\mu_j \geq 0, \forall j \in \mathcal{J}$ and $\nu_{ij} \geq 0, \forall j \in \mathcal{J}, \forall i \in \mathcal{I}_j^\pi$ be the dual variables associated with constraints (3.5) and $\tilde{c}_{ij} \leq 1$ of the problem given in (3.14), respectively. Then, the Lagrangian of (3.14) can be expressed as

$$
\begin{aligned}
L_j\left(\tilde{\mathbf{c}}, \mu, \nu\right) &= \log\left(\sum_{i \in \mathcal{I}_j^\pi} H_{ij}\tilde{c}_{ij} + \iota\right) - \sum_{i \in \mathcal{I}_j^\pi} \nu_{ij}\left(\tilde{c}_{ij} - 1\right) \\
&\quad - \sum_{i \in \mathcal{I}_j^\pi} y_{ij}\tilde{c}_{ij} - \mu_j\left(\sum_{i \in \mathcal{I}_j^\pi} \tilde{c}_{ij} - S_j\right) \\
&= \log\left(\sum_{i \in \mathcal{I}_j^\pi} H_{ij}\tilde{c}_{ij} + \iota\right) \\
&\quad - \sum_{i \in \mathcal{I}_j^\pi}\left(y_{ij} + \mu_j + \nu_{ij}\right)\tilde{c}_{ij} + \mu_j S_j + \sum_{i \in \mathcal{I}_j^\pi} \nu_{ij}
\end{aligned} \tag{3.18}
$$

It is noted that $L_j\left(\tilde{\mathbf{c}}, \mu, \nu\right)$ is a continuous and differentiable function of $\tilde{c}_{ij}$, μ_j, and ν_{ij}. By differentiating $L_j\left(\tilde{\mathbf{c}}, \mu, \nu\right)$ with respect to $\tilde{c}_{ij}$, we have the Karush-Kuhn-Tucker (KKT) conditions as

$$
\frac{\partial L_j}{\partial \tilde{c}_{ij}} = \frac{H_{ij}}{\left(\sum_{i \in \mathcal{I}_j^\pi} H_{ij}\tilde{c}_{ij} + \iota\right)} - \left(y_{ij} + \mu_j + \nu_{ij}\right) \leq 0, \forall i \in \mathcal{I}_j^\pi \tag{3.19}
$$

$$
\tilde{c}_{ij}\left[\frac{H_{ij}}{\left(\sum_{i \in \mathcal{I}_j^\pi} H_{ij}\tilde{c}_{ij} + \iota\right)} - \left(y_{ij} + \mu_j + \nu_{ij}\right)\right] = 0, \forall i \in \mathcal{I}_j^\pi \tag{3.20}
$$

$$
\mu_j\left(\sum_{i \in \mathcal{I}_j^\pi} \tilde{c}_{ij} - S_j\right) = 0, \forall i \in \mathcal{I}_j^\pi \tag{3.21}
$$

$$
\nu_{ij}\left(\tilde{c}_{ij} - 1\right) = 0, \forall i \in \mathcal{I}_j^\pi \tag{3.22}
$$

From (3.19) and (3.20), we obtain an optimal cache allocation for fixed Lagrange multipliers as

$$
\tilde{c}_{ij} = \left[\frac{1}{\left(y_{ij} + \mu_j + \nu_{ij}\right)} - \frac{\iota + \sum_{i' \neq i} H_{i'j}\tilde{c}_{i'j}}{H_{ij}}\right]^+, \tag{3.23}
$$

where $[x]^+ = \max\{x, 0\}$. As $H_{ij} = 0$, $\left[\frac{1}{(y_{ij} + \mu_j + \nu_{ij})} - \frac{\iota + \sum_{i' \neq i} H_{i'j}\tilde{c}_{i'j}}{H_{ij}}\right] \to -\infty$, $\tilde{c}_{ij} = 0$. In other words, since $H_{ij} = 0$, any positive values of $\tilde{c}_{ij}$ do not increase utility but waste physical resources instead. To obtain (3.23), using a gradient-based search, the updated μ_j value is given by

$$
\mu_j^{[t+1]} = \left[\mu_j^{[t]} - \alpha_\mu^{[t]}\left(\sum_{i \in \mathcal{I}_j^\pi} \tilde{c}_{ij} - S_j\right)\right]^+, \tag{3.24}
$$

Algorithm 1: Upper bound algorithm of wireless edge caching

Input: Backhaul price $\{\lambda_j\}$ and caching gain $\{H_{ij}\}$
Output: Relaxed cache placement strategy $\{\tilde{c}_{ij}\}$

1 **begin** Solving problem (3.14)
2   Set prescribed accuracy ζ and maximum number of iteration steps T;
3    $\mu_j \leftarrow \mu_j^{[0]}, \forall j; \quad \nu_{ij} \leftarrow \nu_{ij}^{[0]}, \forall i, j;$ // set initial dual variables
4    $t \leftarrow 1;$// interation step indicator
5   **while** ζ *is not reached* && $t \leq T$ **do**
6   Update $\tilde{c}_{ij}^{[t+1]}$ according to (3.23);
7   Update $\mu_j^{[t+1]}$ according to according to (3.24);
8   Update $\nu_{ij}^{[t+1]}$ according to according to (3.25);
9       $t \leftarrow t + 1;$
10   **end**
11 **end**

---

where $[t]$ is the iteration index and $\alpha_\mu^{[t]}$ are sufficiently small step sizes. Similar to ν_{ij},

$$\nu_{ij}^{[t+1]} = \left[\nu_{ij}^{[t]} - \alpha_\nu^{[t]} \left(\tilde{c}_{ij} - 1 \right) \right]^+, \tag{3.25}$$

where $\alpha_\nu^{[t]}$ are sufficiently small step sizes. We summarize the procedure used to get the optimal solution of the problem (3.14) in Algorithm 1.

The problem (3.15) obviously is a linear problem that can be solved effortlessly by general methods (e.g., interior point method) that are polynomial time algorithms in the worst case. Moreover, since the problem (3.15) is similar to the problem (3.14), even simpler because of the linear objective function, we can use the same idea to solve the problem (3.15).

3.3.3 Rounding methods based on marginal benefits

Recall that we have relaxed the cache placement indicators c_{ij} and video resolution x_{iq} to real values between zero and one instead of binary variables. Thus, we have to recover them to binary values after we get the relaxed solution. The basic idea of the rounding method is that we select the 'best' users to utilize the caching resource and the 'highest' resolution for users under the current resource allocation solution. In this chapter, c_{ij} is recovered to binary based on the corresponding marginal benefit [31, 45].

First, we calculate the marginal benefits of each user as

$$\Omega_{ij}^c = \partial U / \partial \tilde{c}_{ij} = \frac{H_{ij}}{\left(\sum_{i \in \mathcal{I}_j^\pi} H_{ij} \tilde{c}_{ij} + \iota \right)}. \tag{3.26}$$

Then, assuming $\tilde{c}_{ij} \in [0,1]$ is the achieved optimum solution, we can obviously calculate the available amount of cache that can be updated at BS j as

$$N_j = \lfloor \sum_{i \in \mathcal{I}_j^\pi} \tilde{c}_{ij} \rfloor \tag{3.27}$$

where $\lfloor x \rfloor$ means taking the maximum integer value that is less than x. In other words, N_j means the maximum number of content segments that can be cached at each BS j during the scheduling period. N_j users can be selected from $\mathcal{I}_j^\pi$, whose Ω_{ij}^c are larger than other users. Formally, we can separate all users of BS j to two sets $\mathcal{I}_j^{[\pi,+]}$ containing N_j = elements and $\mathcal{I}_j^{[\pi,-]}$ containing the remaining users. $\mathcal{I}_j^{[\pi,+]}$ and $\mathcal{I}_j^{[\pi,-]}$ have following relationship:

$$\max_{i \in \mathcal{I}_j^{[\pi,-]}} \Omega_{ij}^c \leq \min_{i \in \mathcal{I}_j^{[\pi,+]}} \Omega_{ij}^c \tag{3.28}$$

Then, the caching indicator c_{ij} can be recovered by

$$c_{ij} = \begin{cases} 1 & \text{if } i \in \mathcal{I}_j^{[\pi,+]}, \\ 0 & \text{otherwise,} \end{cases} \tag{3.29}$$

The recovering of x_{iq} is easier due to constraints (3.6). By assuming r_{ij}^a is the achieved optimum solution, the method uses the following rule:

$$x_{iq} = \begin{cases} 1, & \text{if } q = \text{argmax}_q \ s_q^v, \text{s.t. } v_q \leq \sum_{i \in \mathcal{J}} r_{ij}^a \\ 0, & \text{otherwise.} \end{cases} \tag{3.30}$$

Then (3.30) is used to find the highest resolution whose required data rate is lower than the provided data rate.

As the two relaxed subproblems have been solved and variables are rounded up to binaries, the solution of the original problem **P0** is obtained. A complete description of the overall proposed scheme is stated in Algorithm 2.

3.3.4 Computational complexity, convergence and optimality

As we mentioned above, the problem formulated in (3.11) is a nonlinear integer optimization problem and is NP-hard without computational efficient algorithms to obtain the optimal solution [41]. An exhaustive searching method can be used to find a solution. However, for a dynamic caching SDWN system with I users and J BSs, the computational complexity of an exhaustive searching method is about $\mathcal{O}((J+1)^I)$ even if we ignore the calculation of spectrum and backhaul allocation, which is tremendously high and unacceptable for practical implementation. The BnB method or dynamic programming can be used to solve this problem, but they are computationally intensive and might not be practical for large-scale problems.

Algorithm 2: The proposed joint allocation algorithm

Input: Wireless network status ($\{a_{lj}\}$ $\{R_{lj}^{max}\}$ $\{\Gamma_{ij}\}$), caching placement ($\{h_{ij}\}$ $\{\pi_{ij}\}$ $\{C_j\}$), and QoE parameters ($\{v_q\}$ S_0 $\{b_{iq}\}$)

Output: $\{r_{lj}^f\}$ $\{r_{ij}^a\}$ $\{c_{ij}\}$ $\{x_{iq}\}$.

1 **begin** Solve the problem **P0**
2   Set prescribed accuracy ϵ and maximum number of iteration steps K;
3      $\lambda_j \leftarrow \lambda_j^{[0]}, \forall j$; // set initial bandwidth prices
4      $k \leftarrow 1$; // interation step indicator
5   **while** ϵ *is not reached* && $k \leq K$ **do**
6   **begin** bandwidth provisioning and video resolution selection
7   Update r_{lj}^f, r_{ij}^a and $\{\tilde{x}_{iq}\}$ by solving the problem (3.15);
8   Round $\{\tilde{x}_{iq}\}$ up to $\{x_{iq}\}$ according to (3.30);
9   **end**
10   **begin** cache placement
11   Update $\tilde{c}_{ij}$ by solving the problem (3.14);
12   Calculate the marginal benefits and available cache according to (3.26) and (3.27);
13   Form $\mathcal{I}_j^{[\pi,-]}$ and $\mathcal{I}_j^{[\pi,+]}$ based on selecting users according to (3.28);
14   Round $\{\tilde{c}_{ij}\}$ up to $\{c_{ij}\}$ according to (3.29);
15   **end**
16   Update $\lambda^{[k+1]}$ according to (3.17);
17         $k \leftarrow k + 1$;
18   **end**
19 **end**

Algorithm 1 is a polynomial time algorithm with the computational complexity of $\mathcal{O}(IJ)$. The computational complexity of solving the problem (3.15) is $\mathcal{O}(\max\{I, J\}J)$, it as a polynomial time algorithm as well. The computational complexity of the proposed dual-decomposition Algorithm 2 is $\mathcal{O}(\max\{I, J\}J)$. The detailed analysis of computational complexity can be found in the following. Obviously, our proposed scheme reduces the computational complexity significantly, compared to the exhaustive searching method.

The complexity of Algorithm 1 can be evaluated as follows. First, the number of variables is IJ and the number of dual variables is $(I + 1)J$ thus the elementary steps needed for calculating $c_{ij}^{[t+1]}$, $\mu_j^{[t+1]}$ and $\nu_{ij}^{[t+1]}$ are $(2I + 1)J$. Second, the maximum loops of iterations for the calculation are T loops. Therefore, as other steps in Algorithm 1 independent from I and J, Algorithm 1 runs in polynomial time with the time complexity of $\mathcal{O}(T(2I+1)J) = \mathcal{O}(IJ)$;

similarly, if Algorithm 1 is running in a distributed manner at each BS. The number of variables and dual variables at each BS is at most I and $(1 + I)$, respectively. Thus, the elementary steps for calculation are $(2I + 1)$, which means the time complexity of each BS is $\mathcal{O}(I)$.

The complexity for solving the problem (3.15) can be evaluated similarly to the above if we use dual methods (e.g., primal-dual interior method). The total number of variables is $QI + IJ + J^2$ and the number of dual variables is $J^2 + J + I + 1 + I + QI$. As a result, the elementary steps needed for calculating those variables at most are $T'(2QI + 2J^2 + IJ + 2I + J + 1)$ where T' is the maximum iterations, which means the time complexity for solving problem (3.15) is $\mathcal{O}(\max\{I, J\}J)$ (a polynomial time algorithm).

In Algorithm 2, the number of bandwidth prises is J, which means J steps are needed to update λ_j. As we explained above, the time complexity for updating $r_{lj}^{f,[k+1]}$, $r_{ij}^{a,[k+1]}$, and $x_{iq}^{[k+1]}$ is $\mathcal{O}(\max\{I, J\}J)$ and for $c_{ij}^{[t+1]}$ is $\mathcal{O}(IJ)$, respectively. Consequently, Algorithm 2 obtains the solution with the the time complexity of $\mathcal{O}(\max\{I, J\}J)$ as the complexities of rounding $\{c_{ij}\}$ and $\{x_{iq}\}$ up are just $\mathcal{O}(IJ)$ and $\mathcal{O}(QI)$ (usually $Q \ll J$).

Either the inner loop in Algorithm 1 or the outer loop in the proposed Algorithm 2 is used to solve a convex problem which is proved to converge to the exact marginal. However, the precise conditions and the initial bandwidth prices used in dual-decomposition cannot be decided before a practical implementation. The optimality cannot be guaranteed since we are solving an NP-hard problem. However, as we can get the upper bound of the proposed problem, we can use it as a replacement for the optimal solution to evaluate our proposed algorithm empirically. Thus, simulation results are used to test the convergence and the optimality of the proposed algorithm (Figure 3.3 in Section 3.4).

3.3.5 Implementation design in SDWNs

To illustrate the utilization of Algorithm 2 into SDWNs, we give an instance presented by a diagram in Figure 3.2.

First, as shown in Figure 3.2, in the control plane, the SDN controller updates the network status according to feedback from BSs, then predicts users' behaviors to calculate $\{\pi_{ij}\}$. By using this information, the SDN controller calls our proposed Algorithm 2 to calculate the provisioned bandwidth and caching strategy for each flow i and BS j. After this, the SDN controller sends the control information to selected BSs and content source (the cloud center or public networks). According to the control information, BSs and other network elements (e.g., potential routers) will update their SDN forwarding tables and perform resource allocation (e.g., cell association and scheduling). In the data plane, flows carrying contents requested by users can pass network elements selected by the control plane to users.

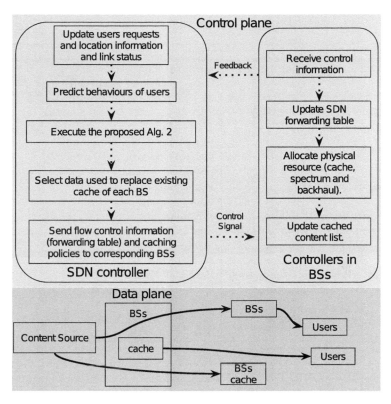

Figure 3.2: The flow diagram in the proposed cache-enabled flow control in SDWNs.

3.4 Simulation results and discussion

Simulation results are presented in this section to demonstrate the performance of the proposed scheme. The simulator is a MATLAB-based system level simulator. Monte Carlo method is used in the simulations. We run the simulations in an X86 desktop computer with quad core CPU (Intel Q8400), 4GB RAM, and the OS of Microsoft Windows 7 SP1. The positions of MBSs and SBSs are fixed. We randomly deploy users in the covered area in each simulation cycle. Average values are taken to reduce randomness effects in the simulations.

The number of available videos is 100,000, with popularity following a Zipf distribution with exponent 0.56, following [17]. In the simulation, we consider a cellular network including 57 BSs that cover 400m-by-400m. Nine BSs are core BSs (macro cells) that connect to the core network and the SDN controller directly. The remaining 48 BSs (small cells) connect to the core networks through these 9 BSs. Besides those, perfect synchronization is assumed in this chapter. The SISO case is considered, but it is attractive to extend it to

Table 3.2: Simulation parameters

Network parameters	value
geographic area	400m-by-400m
number of BSs	9 macro cells, 48 small cells
number of users	30 − 120, 80* is the default
frequency bandwidth (MHz)	20
frequency reuse factor	1
transmission power profile	SISO with maximum power; 49dBm (macro), 20dBm (small)
propagation profile[42]	pathloss:L(distance)=34+40log(distance); lognormal shadowing: 8dB; no fast fading
power density of the noise	-174 dBm/Hz
spectrum efficiency calculation	Shannon bound
backhaul capacity (Mbps)	macro to CN:500; small to macro: 25 − 160, 40*
average trasmission latency (ms)	RAN: 50; backhaul: 60
prediction probability range	low:0-0.5; high:0.5-0.99
total content	10^5
content popularity profile	Zipf distribution with exponent 0.56 [17]
cache size at BSs	300 − 1200, 600*

MIMO in future work. The remaining simulation parameters are summarized in Table. 3.2. The values with $*$ are default values. The information on video streams, such as required data rate and corresponding MV-MOS, is shown in Table 3.3.

In our simulations, we compare five different schemes: the no-caching scheme (baseline (no cache)) that delivers video traffic in wireless networks without caching; the static caching (baseline) scheme (caches are fixed in the research period) considering resource allocation, similar to what is discussed in [17, 1], and the proposed dynamic caching scheme that replaces the caches considering resource allocation and video quality selection. Two scenarios of the content request probability π are tested. One is a high probability ranging from 0.5 to 0.95 and the other is low probability ranging from 0-0.5. The high probability means that user behavior is easier to predict (e.g., watching popular video and tractable paths) and the low probability means the user may quickly change the video and wander in the city.

3.4.1   Algorithm performance

Figure 3.3 demonstrates the evolution of the proposed dual decomposition algorithm for different initial values of λ_j. The iteration step k refers to the main loop iteration of Algorithm 2. At each step we calculate the differences between obtained utility $U^{[k]}$ and the optimum utility U^* by solving the relaxed problem (upper bound) $\tilde{P0}$. In most simulation instances shown in Figure 3.3,

Table 3.3: vMOS of video resolutions[37]

q	resolution	required data rate v_q	resolution vMOS s_q
1	4k or more	25 Mbps	4.9
2	2k	12 Mbps	4.8
3	1080p	10 Mbps	4.5
4	720p	5 Mbps	4
5	480p	2 Mbps	3.6
6	360p	1.5 Mbps	2.8

we see that the proposed algorithm converges to a fixed point. It can be observed that the iterative algorithm converges to the optimal value within ten steps when $\lambda = 0.05$, which means the optimum dynamic caching strategy and the bandwidth provisioning can be achieved within a few iterations. However, Figure 3.3 also suggests that some inappropriate initial values of λ may result in the worst convergence speed. As shown in Figure 3.3, its performance can approach the upper bound with a slight gap, especially after ten steps, which embraces a relatively acceptable solution.

To evaluate the throughput of the proposed scheme, we compare our proposed dynamic caching scheme with two peers. The first one called Baseline (no SDWN) is the proposed caching policy stated in [32] and the second called Baseline (no cache) is the traditional network. As shown in Figure 3.4, both two cases with caching show better performance than the non-caching case. Moreover, by deploying bandwidth provisioning using SDWN, the proposed scheme shows further improvements on the throughput per user compared to other schemes.

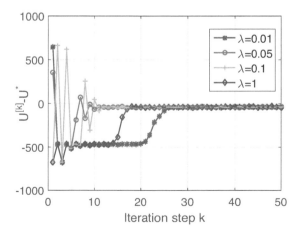

Figure 3.3: Convergence and optimality of the proposed scheme.

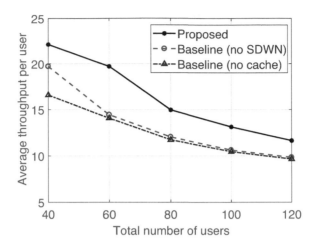

Figure 3.4: Average throughput per user with different network loads.

3.4.2 Network performance

3.4.2.1 Delay

Transmission delay is the key performance metric for video streaming in cache-enabled SDWNs. Users may give up watching the video because of the long waiting time for video buffering. In this subsection, we evaluate the average transmission delay performance and the results are shown in Figures 3.5 and 3.6.

Figure 3.5 shows the cumulative distribution function (CDF) of round trip delay of users with different caching schemes and prediction accuracy,

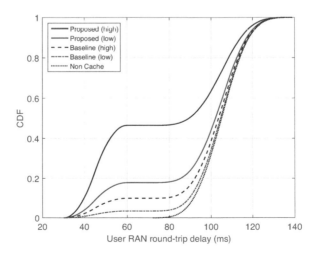

Figure 3.5: CDF of round trip delay (backhaul and RAN only) of users.

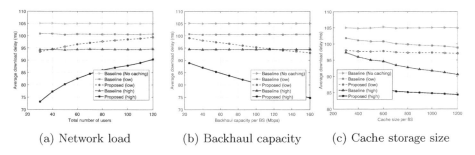

(a) Network load	(b) Backhaul capacity	(c) Cache storage size

Figure 3.6: Average mobile network delay with different network setups.

respectively. The CDFs for dynamic and static caching all improve significantly at delay reduction compared to the no in-network caching case, showing a maximum 2.5x gain, in both low and high probability settings. Specifically, the proposed proactive caching boosts the performance of network delay deduction the most when we can keep caching contents with high probability to be requested. Almost 50% of users experience delay less than 60 ms because their backhaul latency is eliminated by the cache. Moreover, we can see even with the low probability setting the performance surpasses the static caching strategy, which implies that the proposed caching scheme should be deployed no matter the probability. It should be noted that flat curves appeared in CDFs (between 60 ms to 80 ms) due to the elimination of backhaul latency brought by deploying in-network caching.

The results shown in Figure 3.6 compare the average delay performance among different caching schemes and prediction accuracies. In Figure 3.6a and Figure 3.6b, note that the average delay of users can be further reduced by increasing available physical resources such as available cache space and/or the backhaul bandwidth of each BS because more data are going to be cached when we have more resources. It is observed that the proposed dynamic caching scheme with high probability always achieves the lowest while the proposed dynamic caching scheme with low probability is also better than the static caching scheme. However, when the probability is not high, increasing backhaul resources or cache resources has a tiny effect on the delay reduction. Moreover, it can be seen in Figure 3.6b that the proposed schemes with low probability surpass the baseline when the backhaul capacity is larger than 90 Mbps. The probable reason is that the bottleneck here is the behavior of users instead of the physical resource. As shown in Figure 3.6c, the increase of users degrades the performance of the network, because it implies that more physical resources are put to real-time flows rather than caching data. This results in fewer hitting events that can reduce delay. It is interesting to observe in Figure 3.6c that the increase of traffic load has very little effect on the cases where no dynamic caching is deployed. The intuitive reason is that static caching is independent of traffic load.

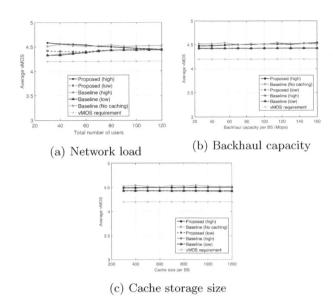

(a) Network load   (b) Backhaul capacity

(c) Cache storage size

Figure 3.7: Average vMoS with different network setups.

3.4.2.2 QoE guarantee

In addition to the mean delay, it is also necessary to analyze the mobile MV-MOS of our proposed scheme. As shown in Figure 3.7a, our proposed scheme satisfies the QoE requirements. Specifically, by putting more users into the network, the vMOS requirements are guaranteed, even there is a slight decrease. Obviously, as shown in the other two figures, the boost of physical resources provides the better situation where degradation would not happen. Therefore, as mentioned in Section 3.3.1, the proposed proactive caching does not undermine the video quality.

3.4.3 Utilization

In this subsection, the utilization of caching resources and backhaul are tested in terms of hitting events, hitting rate and backhaul load.

3.4.3.1 Caching resources

The hit ratio is widely adopted as the performance metric to evaluate caching mechanisms [7]. Since our first baseline does not consider an in-network caching scheme, consequently, we omit the no-caching scheme in this subsection and illustrate the average hit ratio of the cached contents of the other four schemes in our simulations in Figure 3.8. Due to dynamic caching operations, the proposed scheme improves the average hit ratio performance by around 30% compared to the fixed caching scheme when the probability is

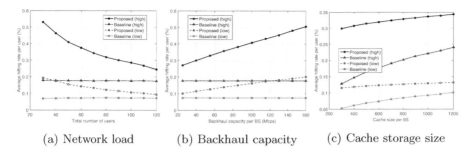

(a) Network load (b) Backhaul capacity (c) Cache storage size

Figure 3.8: Average hit ratio with different network setups.

high, and by around 10% at most if the probability is lower. In Figure 3.8a, it is observed that the hitting ratio of both proposed cases decrease with the total number of users because more users increase the load of backhaul used for streaming videos.

Another performance measurement metric, total hitting events, also can be used to evaluate the utilization of cache resources at each BS. Figure 3.9 shows the average total hitting within one hour per BS. Obviously, our proposed schemes show better performance than passive caching schemes. Moreover, with increasing the number of users in the networks, the total cache hitting of fixed caching is increased due to the fact that the larger number of users boosts the opportunities. Besides network load, with the increase of backhaul capacity of BSs and the cache capacity of each BS, the average hitting events increase significantly by deploying dynamic caching. The reason is the same mentioned in the analysis of delay.

3.4.3.2 Backhaul resource

In this experience, we compare the (normalized) backhaul traffic load of video streaming with different schemes in Figure 3.10 with different parameters. In Figure 3.10, we can see, no-caching scheme takes the most backhaul load while all other four caching schemes can alleviate backhaul. Specifically, with varying

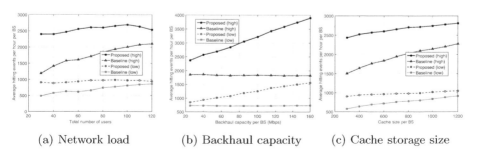

(a) Network load (b) Backhaul capacity (c) Cache storage size

Figure 3.9: Average hitting events with different network setups.

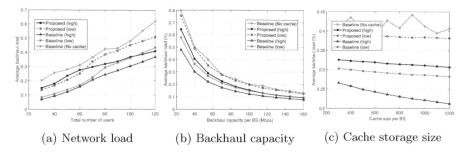

(a) Network load   (b) Backhaul capacity   (c) Cache storage size

Figure 3.10: Average backhaul load with different network setups.

network status (network load, backhaul capacity and cache size), dynamic caching schemes cost more backhaul as they replace the caches more frequently than fixed caching schemes. Furthermore,the two caching cases (dynamic and fixed) of high probability save more backhaul. Apparently, higher number of users leads the higher backhaul load shown in Figure 3.10a and more physical resources give more flexible backhaul pressure as shown in Figure 3.10b and Figure 3.10c.

3.5 Conclusions and future work

In this chapter, we jointly studied radio resource allocation, dynamic caching and adaptive video resolution selection for cache-enabled SDWNs. We proposed a polynomial time algorithm to solve the joint problem. We took dynamic caching and adaptive video resolution selection into account in the flow control problem in SDWNs and formulated a joint problem. Then, we transferred this problem by relaxation to a convex problem that can be solved efficiently, and an algorithm was designed. Simulation results were presented to show that the performance of our proposed scheme can improve the QoE in terms of delay and video quality as well as efficiency of physical resource utilization. Future work is in progress to consider mobile edge computing in the proposed framework.

References

[1] *Cisco Visual Networking Index: Global Mobile Data Traffic Forecast Update 2016–2021 White Paper.* Accessed on Aug. 02, 2017. [Online]. Available: http://www.cisco.com/c/en/us/solutions/collateral/serviceprovider/visual-networking-index-vni/complete-white-paper-c11-481360.pdf

[2] B. A. A. Nunes, M. Mendonca, X.-N. Nguyen, K. Obraczka, and T.

Turletti, "A survey of software-defined networking: Past, present, and future of programmable networks," *IEEE Commun. Surveys Tuts.*, vol. 16, no. 3, pp. 1617–1634, 2014.

[3] T. Chen, M. Matinmikko, X. Chen, X. Zhou, and P. Ahokangas, "Software defined mobile networks: Concept, survey, and research directions," *IEEE Commun. Mag.*, vol. 53, no. 11, pp. 126–133, Nov. 2015.

[4] R. Q. Hu and Y. Qian, "An energy efficient and spectrum efficient wireless heterogeneous network framework for 5G systems," *IEEE Commun. Mag.*, vol. 52, no. 5, pp. 94–101, May 2014.

[5] Y. Xu, R. Q. Hu, Y. Qian, and T. Znati, "Video quality-based spectral and energy efficient mobile association in heterogeneous wireless networks," *IEEE Trans. Commun.*, vol. 64, no. 2, pp. 805–817, Feb. 2016.

[6] Y. Cai, F. R. Yu, C. Liang, B. Sun, and Q. Yan, "Software-defined device-to-device (D2D) communications in virtual wireless networks with imperfect network state information (NSI)," *IEEE Trans. Veh. Tech.*, vol. 65, no. 9, pp. 7349–7360, Sep. 2016.

[7] P. Si, H. Yue, Y. Zhang, and Y. Fang, "Spectrum management for proactive video caching in information-centric cognitive radio networks," *IEEE J. Sel. Areas Commun.*, vol. 34, no. 8, pp. 2247–2259, Aug. 2016.

[8] C. Liang, F. R. Yu, and X. Zhang, "Information centric network function virtualization over 5G mobile wireless networks," *IEEE Netw.*, vol. 29, no. 3, pp. 68–74, May 2015.

[9] W.-C. Liao, M. Hong, H. Farmanbar, X. Li, Z.-Q. Luo, and H. Zhang, "Min flow rate maximization for software defined radio access networks," *IEEE J. Sel. Areas Commun.*, vol. 32, no. 6, pp. 1282–1294, Jun. 2014.

[10] B. Ahlgren, C. Dannewitz, C. Imbrenda, D. Kutscher, and B. Ohlman, "A survey of information-centric networking," *IEEE Commun. Mag.*, vol. 50, no. 7, pp. 26–36, Jul. 2012.

[11] D. Liu, B. Chen, C. Yang, and A. F. Molisch, "Caching at the wireless edge: Design aspects, challenges, and future directions," *IEEE Commun. Mag.*, vol. 54, no. 9, pp. 22–28, Sep. 2016.

[12] J. Li, Y. Chen, Z. Lin, W. Chen, B. Vucetic, and L. Hanzo, "Distributed caching for data dissemination in the downlink of heterogeneous networks," *IEEE Trans. Commun.*, vol. 63, no. 10, pp. 3553–3568, Oct. 2015.

[13] M. Tao, E. Chen, H. Zhou, and W. Yu, "Content-centric sparse multicast beamforming for cache-enabled cloud RAN," *IEEE Trans. Wireless Commun.*, vol. 15, no. 9, pp. 6118–6131, Sep. 2016.

[14] H. A. Pedersen and S. Dey, "Enhancing mobile video capacity and quality using rate adaptation, RAN caching and processing," *ACM/IEEE Trans. Netw.*, vol. 24, no. 2, pp. 996–1010, Apr. 2016.

[15] Y. Chen, K. Wu, and Q. Zhang, "From QoS to QoE: A tutorial on video quality assessment," *IEEE Commun. Surveys Tuts.* , vol. 17, no. 2, pp. 1126–1165, 2015.

[16] N. Dao, H. Zhang, H. Farmanbar, X. Li, and A. Callard, "Handling real-time video traffic in software-defined radio access networks," in *Proc. IEEE ICC Workshops*, Jun. 2015, pp. 191–196.

[17] N. Golrezaei, K. Shanmugam, A. G. Dimakis, A. F. Molisch, and G. Caire, "Femtocaching: Wireless video content delivery through distributed caching helpers," in *Proc. IEEE INFOCOM*, Mar. 2012, pp. 1107–1115.

[18] H. Farmanbar and H. Zhang, "Traffic engineering for software-defined radio access networks," in *Proc. IEEE Netw. Oper. Manag. Symp. (NOMS)*, May 2014, pp. 1–7.

[19] M. S. El Bamby, M. Bennis, W. Saad, and M. Latva-aho, "Contentaware user clustering and caching in wireless small cell networks," in *Proc. 11th Int. Symp. Wireless Commun. Syst. (ISWCS)*, Aug. 2014, pp. 945–949.

[20] A. Liu and V. K. N. Lau, "Mixed-timescale precoding and cache control in cached mimo interference network," *IEEE Trans. Signal Process.*, vol. 61, no. 24, pp. 6320–6332, Dec. 2013.

[21] B. Dai and W. Yu, "Sparse beamforming and user-centric clustering for downlink cloud radio access network," *IEEE Access*, vol. 2, pp. 1326–1339, Oct. 2014.

[22] A. Abboud, E. Batu, K. Hamidouche, and M. Debbah, "Distributed caching in 5G networks: An alternating direction method of multipliers approach," in *Proc. IEEE Workshop Signal Process. Adv. Wireless Commun. (SPAWC)*, Jul. 2015, pp. 171–175.

[23] C. Liang and F. R. Yu, "Bandwidth provisioning in cache-enabled software-defined mobile networks: A robust optimization approach," in *Proc. IEEE Veh. Tech. Conf. (VTC-Fall)*, Sep. 2016, pp. 1–5.

[24] K. Poularakis, G. Iosifidis, and L. Tassiulas, "Joint caching and base station activation for green heterogeneous cellular networks," in *Proc. IEEE Int. Conf. Commun. (ICC)*, Jun. 2015, pp. 3364–3369.

[25] V. A. Siris, X. Vasilakos, and G. C. Polyzos, "Efficient proactive caching for supporting seamless mobility," in *Proc. Proc. IEEE 15th Int. Symp. World Wireless, Mobile Multimedia Netw. (WoWMoM)*, Jun. 2014, pp. 1 6.

[26] M. Seufert, S. Egger, M. Slanina, T. Zinner, T. Hofeld, and P. Tran-Gia, "A survey on quality of experience of HTTP adaptive streaming," *IEEE Commun. Surveys Tuts.* , vol. 17, no. 1, pp. 469–492, 2015.

[27] H. Ahlehagh and S. Dey, "Video-aware scheduling and caching in the radio access network," *IEEE/ACM Trans. Netw.*, vol. 22, no. 5, pp. 1444–1462, Oct. 2014.

[28] J. Zhu, J. He, H. Zhou, and B. Zhao, "EPCache: In-network video caching for LTE core networks," in *Proc. Int. Conf. Wireless Commun. Signal Process. (WCSP)*, Oct. 2013, pp. 1–6.

[29] A. Liu and V. K. N. Lau, "Exploiting base station caching in MIMO cellular networks: Opportunistic cooperation for video streaming," *IEEE Trans. Signal Process.*, vol. 63, no. 1, pp. 57–69, Jan. 2015.

[30] R. Yu et al., "Enhancing software-defined RAN with collaborative caching and scalable video coding," in *Proc. IEEE Int. Conf. Commun. (ICC)*, May 2016, pp. 1–6.

[31] L. Ma, F. Yu, V. C. M. Leung, and T. Randhawa, "A new method to support UMTS/WLAN vertical handover using SCTP," *IEEE Wireless Commun.*, vol. 11, no. 4, pp. 44–51, Aug. 2004.

[32] C. Liang, F. R. Yu, H. Yao, and Z. Han, "Virtual resource allocation in information-centric wireless networks with virtualization," *IEEE Trans. Veh. Technol.*, vol. 65, no. 12, pp. 9902–9914, Dec. 2016.

[33] D. Niyato, D. I. Kim, P. Wang, and M. Bennis, "Joint admission control and content caching policy for energy harvesting access points," in *Proc. IEEE Int. Conf. Commun. (ICC)*, May 2016, pp. 1–6.

[34] J. Qiao, Y. He, and X. S. Shen, "Proactive caching for mobile video streaming in millimeter wave 5G networks," *IEEE Trans. Wireless Commun.*, vol. 15, no. 10, pp. 7187–7198, Oct. 2016.

[35] F. Yu and V. C. M. Leung, "Mobility-based predictive call admission control and bandwidth reservation in wireless cellular networks," in *Proc. IEEE INFOCOM'01*, Anchorage, AK, USA, Apr. 2001, pp. 518–526.

[36] Y. He, F. R. Yu, N. Zhao, H. Yin, H. Yao, and R. C. Qiu, "Big data analytics in mobile cellular networks," *IEEE Access*, vol. 4, pp. 1985–1996, Mar. 2016.

[37] Y. He, C. Liang, F. R. Yu, N. Zhao, and H. Yin, "Optimization of cache-enabled opportunistic interference alignment wireless networks: A big data deep reinforcement learning approach," in *Proc. IEEE Int. Conf. Commun. (ICC)*, Paris, France, Jun. 2017, pp. 1–6.

[38] D. Schoolar. (2015) Whitepaper: Mobile Video Requires Performance and Measurement Standards. [Online]. Available: http://wwwfile.huawei.com/

[39] S. Singh, O. Oyman, A. Papathanassiou, D. Chatterjee, and J. G. Andrews, "Video capacity and QoE enhancements over LTE," in *Proc. IEEE Int. Conf. Commun. (ICC)*, Jun. 2012, pp. 7071–7076.

[40] A. El Essaili, D. Schroeder, E. Steinbach, D. Staehle, and M. Shehada, "QoE-based traffic and resource management for adaptive HTTP video delivery in LTE," *IEEE Trans. Circuits Syst. Video Technol.*, vol. 25, no. 6, pp. 988–1001, Jun. 2015.

[41] D. Bethanabhotla, O. Y. Bursalioglu, H. C. Papadopoulos, and G. Caire, "Optimal user-cell association for massive MIMO wireless networks," *IEEE Trans. Wireless Commun.*, vol. 15, no. 3, pp. 1835–1850, Mar. 2016.

[42] G. Li and H. Liu, "Downlink radio resource allocation for multi-cell OFDMA system," *IEEE Trans. Wireless Commun.*, vol. 5, no. 12, pp. 3451–3459, Dec. 2006.

[43] Q. Ye, B. Rong, Y. Chen, M. Al-Shalash, C. Caramanis, and J. G. Andrews, "User association for load balancing in heterogeneous cellular networks," *IEEE Trans. Wireless Commun.*, vol. 12, no. 6, pp. 2706–2716, Jun. 2013.

[44] F. Yu and V. Krishnamurthy, "Effective bandwidth of multimedia traffic in packet wireless CDMA networks with LMMSE receivers: A crosslayer perspective," *IEEE Trans. Wireless Commun.*, vol. 5, no. 3, pp. 525–530, Mar. 2006.

[45] S. Boyd, L. Xiao, A. Mutapcic, and J. Mattingley, "Notes on decomposition methods," Stanford Univ., Stanford, CA, USA, Tech. Rep. EE364B, 2003. [Online]. Available: https://stanford.edu/class/ee364b/lectures/decomposition_notes.pdf

[46] G. Liu, F. R. Yu, H. Ji, and V. C. M. Leung, "Energy-efficient resource allocation in cellular networks with shared full-duplex relaying," *IEEE Trans. Veh. Tech.*, vol. 64, no. 8, pp. 3711–3724, Aug. 2015.

Chapter 4

Resource Allocation for 3C-Enabled HetNets

In order to better accommodate the dramatically increasing demand for data caching and computing services, storage and computation capabilities should be placed on some of the intermediate nodes within the network, therefore increasing data throughput and reducing network operation cost. In this chapter, we design a novel information-centric heterogeneous networks framework aiming at enabling content caching and computing. Furthermore, due to the virtualization of the whole system, the communication, computing and caching (3C) resources can be shared among all users associated with different virtual service providers. We formulate the virtual resource allocation strategy as a joint optimization problem, where the gains of not only virtualization but also caching and computing are taken into consideration in the proposed information-centric heterogeneous networks virtualization architecture. In addition, a distributed algorithm based on the alternating direction method of multipliers is adopted in order to solve the formulated problem. Since each base station only needs to solve its own problem without exchange of channel state information by using the distributed algorithm, the computational complexity and signaling overhead can be greatly reduced. Finally, extensive simulations are presented to show the effectiveness of the proposed scheme under different system parameters.

4.1 Introduction

According to the investigation conducted in [1], global video traffic is assumed to dominate the Internet by making up 80% of total Internet usage by 2018. However, it is still extremely difficult to satisfy the soaring demand of resource-constrained mobile devices for video consumption. Since heterogeneous devices, such as TV, PC, tablet, and smartphone, request different kinds of video formats, resolutions, and bitrates, existing video contents may need to be transformed to fit the network condition and the usage of different mobile device [2]. Therefore, transcoding technology is necessary for transforming the current video version into a suitable one, which can be played and matches the screen size of the devices [3]. However, such a transcoding procedure is computationally intensive, so that it can hardly be executed on mobile devices with limited resources. As such, a novel computing platform is desirable.

Mobile edge computing (MEC) is recognized as a promising paradigm in next generation wireless networks, enabling cloud-computing capabilities in close proximity to mobile devices [4, 5]. With the physical proximity, MEC realizes a low-latency connection to a large-scale resource-rich computing infrastructure by offloading the computation task to an adjacent computing server/cluster instead of relying on a remote cloud [6]. With MEC, the need for fast interactive response can be met within the network edge. Therefore, MEC is envisioned to provide computation services for mobile devices at any time and anywhere by endowing radio access networks (RANs) with powerful computing capabilities [5].

The prodigious amount of videos and the wide variety of video versions will certainly result in a large-scale distribution of video contents calling for tremendous resources. Therefore, it is essential to have computing and storage resources at some of the intermediate nodes within the network [1, 7], in order to jointly transcode/compute and cache them, meanwhile reducing the backhaul consumption of popular contents.

Recent advances in *information-centric networking* (ICN) and wireless network virtualization can help efficient distribution of video contents in wireless networks. ICN is capable of converting the sender-driven end-to-end network paradigm into the receiver-driven content retrieval paradigm [8]. ICN is characterized by receiver-driven information-level delivery and *in-network caching*. Compared to the traditional network paradigms with a general lack of content distribution information, ICN can reduce the backhaul cost of popular contents, increase the delivery probability of contents to mobile users, and support a highly efficient and scalable content retrieval [9]. For next generation wireless networks, ICN-based air caching is also recognized as a promising technique [10].

Wireless network virtualization has also attracted great attention, since it is able to significantly reduce capital expenses and operational expenses of wireless access networks as well as core networks [11]. Through virtualization, wireless network infrastructure can be decoupled from the provided services, and various users with different service requirements can dynamically share the same infrastructure, thereby maximizing system utilization [12, 13].

It is a new trend to integrate wireless network virtualization with ICN technique in next generation wireless cellular networks, since such an integration can further improve end-to-end network performance [10]. Specifically, ICN virtualization enables the sharing of the infrastructures and the contents among users from different service providers (SPs), thus promoting the gains of not only in-network caching but also virtualization [12, 13].

Thanks to ICN virtualization and MEC, computing and caching functions can be offered within close proximity to mobile devices. However, although some seminal works have been done on ICN virtualization and MEC, these two areas have been addressed separately. Thus, how to combine these two techniques, and efficiently allocate the limited resources to jointly optimize the utilities of computing, caching, and communication, remain to be urgent issues.

In this chapter, we investigate information-centric virtual heterogeneous networks (HetNets) with in-network caching and MEC. The distinctive features of this chapter are as follows:

1. We design a novel information-centric HetNets framework aiming at enabling content caching and computing. Moreover, the whole system is virtualized, in which the resources of communication, computing, and caching can be shared among users from different virtual networks.

2. In this framework, we formulate the virtual resource allocation strategy as a joint optimization problem, where the gains of not only virtualization but also caching and computing are taken into consideration in the proposed information-centric HetNets virtualization architecture.

3. A distributed algorithm, based on the alternating direction method of multipliers (ADMM) [14], is presented to solve the formulated problem. Since each infrastructure provider (InP) only needs to solve its own problem without exchange of channel state information (CSI) by using this distributed algorithm, computational complexity and signaling overhead can be greatly reduced.

4. Extensive simulations are conducted with different system parameters to investigate different aspects of the virtual system and to evaluate the performance of the proposed algorithm. The results show that the total utility of a mobile virtual network operator (MVNO) can be significantly improved with a fast convergence rate in the proposed scheme.

4.2 Architecture overview

In this section, we introduce wireless network virtualization, ICN, and MEC. Then, we present a virtualized HetNets framework with in-network caching and MEC functions.

4.2.1 Wireless network virtualization

In traditional physical networks, such as cellular and core networks, the whole system is designed to support different kinds of services. Compared to traditional physical networks, a virtual wireless network with a MVNO can be designed to focus on only one type of service, such as streaming video or online games, which will certainly result in a better user experience.

The logical roles after virtualization can be identified as infrastructure provider (InP), service provider (SP), and mobile virtual network operator (MVNO) [15], the main functions of which are described as follows:

- InP: operates the physical network infrastructures, and owns radio resources of physical substrate networks, including licensed spectrum and backhaul.

- SP: leases and manages the virtual resources to mobile users for providing end-to-end services.

- MVNO: leases the radio resources from InPs, generates the virtual radio resources according to the requests of users belonging to different SPs, and operates and leases the virtual resources to SPs.

With the trend of decoupling network services from specific network hardware, wireless network virtualization is expected to be applied in future mobile network frameworks, such as software-defined wireless networks [16, 17, 18, 19].

4.2.2 Information-centric networking

In ICN, mobile users care much about what the content is rather than the location of content. All contents are named so that they can be independent of the locations of the nodes possessing the contents. The user who requests one content does not need to gather information about the content holders. Instead, according to the receiver-driven principle, the communication link is set up by the user to the provider, and the required content is transferred through the reverse path. In other words, the primary purpose of ICN is to find, deliver, and disseminate information effectively and efficiently, rather than to make communication between users and content holders available and maintained. The match of the requested content instead of the findability of the content holder thus dictates the establishment of communication.

Another main function of ICN is in-network caching, with which nodes are capable of caching content with a higher popularity passing through them based on the remaining caching spaces. Afterwards, if some users request contents that have already been cached, contents can be delivered directly to the users. Via such a mechanism, the backhaul consumptions of the popular contents are decreased, and meanwhile the delivery probabilities of contents to mobile users are increased.

4.2.3 Mobile edge computing

In September 2014, the MEC Industry Initiative introduced a reference architecture to declare the challenges that need to be bypassed while implementing the architecture of MEC [19]. The principal concern of MEC is to provide cloud computing capability and information technology (IT) within RAN in close proximity to mobile users. In general, if a mobile device requires a much higher computation rate, or when the time (or energy consumption) for executing a given task on the mobile device locally is much longer (or higher) than the response time (or energy consumption) of offloading and computing the task onto an MEC server, computation offloading should be performed. Through exploiting agility, context, proximity, and speed, MEC is able to transform mobile base stations (BSs) into intelligent service hubs for creating a new value chain and stimulating revenue generation. Meanwhile, MEC is expected to enable the network to support a wide range of new services and applications [21]. In the case of LTE, as shown in [22], the MEC server is integrated into the eNodeB directly.

In this chapter, we assume that the MEC server is integrated into the BS, as shown in Figure 4.1. The BS can provide computation services for mobile devices within its working range. The mobile devices can run mobile applications locally, or offload some computation tasks to the BS. In our business model, compared to the conventional way of accessing distant cloud infrastructure, the mobile device is capable of executing the computation task at the BS as long as the BS is available between the mobile device and MEC server. This enables a much faster computation rate along with a lower operation cost compared with that of executing the tasks at the distant cloud.

4.2.4 3C-enabled virtualized HetNets

Figure 4.2 illustrates the proposed framework of information-centric HetNets virtualization with in-network caching and MEC functions. In this model, two virtual wireless networks, which are embedded in the physical substrate networks, are established according to the types of heterogeneous services. Considering that mobile users may have some computation tasks (e.g., transcode video) that need to be offloaded, it is beneficial for the whole system to have a computing resource at each BS. The computing resource can also be virtualized at each BS. Each mobile user can connect to virtual wireless networks logically, and subscribe to the required services from these virtual networks,

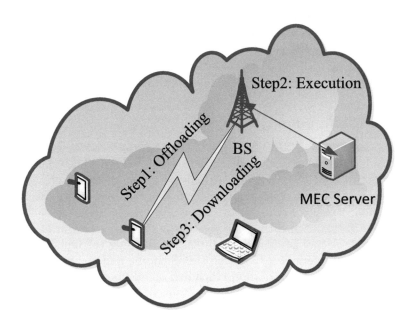

Figure 4.1: A network framework integrated with an MEC server.

while actually they connect to physical networks. It is assumed that the whole virtualization process is realized and controlled by a virtual wireless network controller.

As mentioned above, virtual radio resources are generated according to requests of SPs by MVNO, since the quality-of-service (QoS) requirements may be different for each mobile user. The two kinds of virtual networks shown in Figure 4.2 can be further illustrated as follows:

■ In the left virtual network (i.e, SP1), the in-network caching function is integrated and cached contents can be shared among users. Some computation tasks like video transcoding are better executed in SP1. Specifically, for adapting complicated network conditions and the usage of various mobile devices, each video content has to be transcoded into multiple versions, which will certainly result in a large-scale distribution of video content calling for tremendous bandwidth resources, especially during the peak time. In such a situation, SP1 offers an opportunity to cache popular contents before or after the execution of the computation tasks, which will significantly reduce the operation cost for delivering video contents.

■ The virtual network on the right (i.e., SP2) is created without a caching

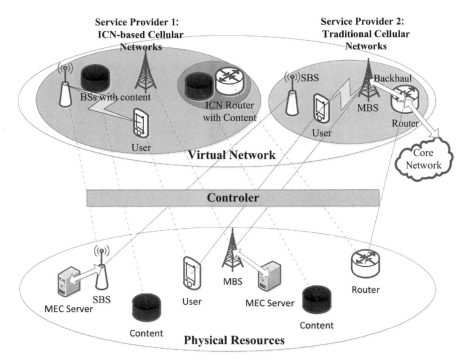

Figure 4.2: Information-centric wireless virtualized HetNets with in-network caching and MEC.

function. Some users, who want to compute a task such as face recognition, prefer to access such a virtual network, because the contents of these kinds of computation tasks may be private and with very low reuse probability.

In summary, there are two important issues that need to be solved in our proposed information-centric HetNets virtualization architecture, which are the virtual resource allocation scheme, including computation resource and communication resource, and the in-network caching strategy. The virtualization procedure performed by MVNO is regarded as the matching between physical resources and virtual resources. MVNO needs to allocate appropriate physical resources of InPs to each SP, so that each mobile user is able to offload computation task to the powerful computing server. Moreover, the in-network caching strategy is designed to decide whether or not to cache contents before or after the computation at BSs based on potential benefits, such as release of redundant traffic and decrease of delay. In other words, the problem is how to design an efficient scheme to jointly solve these problems in the proposed framework, which will be formulated as an optimization problem in the following.

4.3 Virtualized multi-resources allocation

In this section, the virtual resource allocation scheme with in-network caching and MEC techniques is formulated to maximize total system utility.

4.3.1 System model

In this subsection, we present the system model for virtual HetNets, in-network caching, and MEC.

4.3.1.1 Virtual heterogeneous networks model

Considering that the HetNets with multiple small cell base station (SBSs) coexist within the coverage of a macro base station (MBS) for serving multiple users, let $\mathcal{N}_s$ be the sets of SBSs, and $\mathcal{N} = \{0\} \cup \mathcal{N}_s = \{0, 1, ..., N\}$ and $\mathcal{U} = \{1, ..., U\}$ be the sets of all BSs and users, respectively. It is assumed that each BS belongs to a different InP, and the licensed spectrum of each InP is orthogonal so that there is no interference among them. In addition, let $\mathcal{S} = \{1, ..., S\}$ be the set of SPs. For each SP s, each assigned user is denoted by u_s, and $\mathcal{U}_s$ is the set of users belonging to SP s, where $\mathcal{U} = \cup_s \mathcal{U}_s$ and $\mathcal{U}_s \cap \mathcal{U}_{s'} = \phi, \forall s' \neq s$.

In our business model, MVNO leases radio resources (e.g., spectrum) and backhaul bandwidth (e.g., data rate) from InPs, and slices them to virtual SPs. On the revenue side, MVNO charges users a virtual network access fee of α_{u_s} per bps. Users who have already paid the fee can access the virtual network for offloading their computation task. The fee for user u_s to compute the task at BSs is defined as ϕ_{u_s} per bps. Since the contents of the computation tasks may have the potential benefit to be cached, the backhaul cost, paid by MVNO and defined as γ_n per bps, can be saved when users call for the contents which have already been cached at BS n. On the spending side, MVNO needs to dynamically pay for the usage of spectrum to InPs, which is defined as β_n per Hz. Furthermore, MVNO also needs to pay the computation fee and caching fee to InPs, once there is a computation task that needs to be executed at the MEC server or the contents before and after the computation are valuable to be cached at BSs. The unit price of computation energy at BS n is defined as ψ_n per J. The prices per unit of space to cache the contents before and after the computation at BS n are denoted by $\Psi_{z_{u_s}}^n$ and $\Psi_{z'_{u_s}}^n$, where z_{u_s} and z'_{u_s} represent the contents before and after the computation. Table 4.1 gives the key parameters in the system model.

4.3.1.2 Computing model

Assume that each user has a computation task to be completed with a certain requirement of computation rate. Let $a_{u_s,n}$ denote the association indicator, where $a_{u_s,n} = 1$ means that user u_s associates with BS n to compute the offloading task; otherwise $a_{u_s,n} = 0$. Practically, each user can associate to

Table 4.1: Key Parameters in the System Model

Notation	*Definition*
$\mathcal{N}_s$	The set of SBSs.
$\mathcal{N}$	The set of all the BSs, and $\mathcal{N} = \{0\} \cup \mathcal{N}_s = \{0, 1, ..., N\}$.
$\mathcal{S} = \{1, ..., S\}$	The set of SPs.
u_s	The allocated user u for SP s.
$\mathcal{U}_s$	The set of users belonging to SP s, and $\mathcal{U}_s \cap \mathcal{U}_{s'} = \phi$, $\forall s' \neq s$.
$\mathcal{U} = \{1, ..., U\}$	The set of all the users, and $\mathcal{U} = \cup_s \mathcal{U}_s$.
α_{u_s}	The access fee from user u_s, charged by MVNO, is defined as α_{u_s} per bps.
ϕ_{u_s}	The computation fee from user u_s, charged by MVNO, is defined as ϕ_{u_s} per bps.
γ_n	The backhaul cost of BS n, paid by MVNO, is defined as γ_n per bps.
β_n	The usage cost of spectrum of BS n, paid by MVNO, is defined as β_n per Hz.
ψ_n	The computation fee to compute the task at BS n, paid by MVNO, is defined as ψ_n per J.
$\Psi^n_{z_{u_s}}$	The storage fee to cache the content z_{u_s} at BS n, paid by MVNO, is defined as $\Psi^n_{z_{u_s}}$ per byte.
$\Psi^n_{z'_{u_s}}$	The storage fee to cache the content z'_{u_s} at BS n, paid by MVNO, is defined as $\Psi^n_{z'_{u_s}}$ per byte.
B_n	The spectrum bandwidth allocated to BS n.
$R^{\mathrm{cm}}_{u_s}$	The communication rate requirement of user u_s in the corresponding QoS class.
$R^{\mathrm{cp}}_{u_s}$	The computation rate requirement of user u_s.
D_n	The total workload at BS n.
Z_n	The storage space of BS n.
σ	The power spectral density (PSD) of the additive white Gaussian noise (AWGN).
q_{u_s}	The transmit power of user u_s.
$h_{u_s,n}$	The channel gain between user u_s and BS n.
z_{u_s}	The size of user u_s's content before computation.
z'_{u_s}	The size of user u_s's content after computation.
$T_{z_{u_s}}$	The time duration for downloading the content z_{u_s} through backhaul.
c_{u_s}	The computing ability required for accomplishing the task of user u_s.
e_n	The energy consumption for one CPU cycle at BS n.
$f_{u_s,n}$	The computation capability of BS n assigned to user u_s.

only one BS; thus

$$\sum_{n \in \mathcal{N}} a_{u_s,n} = 1, \forall s \in \mathcal{S}, u_s \in \mathcal{U}_s. \tag{4.1}$$

$b_{u_s,n}$ denotes the allocated bandwidth from BS n to user u_s, and we have

$$\sum_{s \in \mathcal{S}} \sum_{u_s \in \mathcal{U}_s} a_{u_s,n} b_{u_s,n} \leq B_n, \forall n \in \mathcal{N} \tag{4.2}$$

where B_n is used to denote the spectrum bandwidth allocated to BS n. In order to ensure the data rate requirements of each user, we have

$$\sum_{n \in \mathcal{N}} a_{u_s,n} b_{u_s,n} r_{u_s,n} \geq R_{u_s}^{\mathrm{cm}}, \forall s \in \mathcal{S}, u_s \in \mathcal{U}_s. \tag{4.3}$$

where $R_{u_s}^{\mathrm{cm}}$ is user u_s's communication rate requirement in the corresponding QoS class.

According to the Shannon bound, $r_{u_s,n}$, the achievable spectrum efficiency of user u_s associating with BS n, can be easily obtained as $r_{u_s,n} = \log_2(1 + \frac{q_{u_s} h_{u_s,n}}{\sigma})$. Here, σ denotes the power spectrum density (PSD) of the additive white Gaussian noise (AWGN), q_{u_s} is the transmission power of user u_s, and $h_{u_s,n}$ represents the large-scale channel gain including pathloss and shadowing. To mitigate cross-tier interference, we adopt a split-spectrum approach where the same spectrum is split between the macro cell and its nearby interfering small cells [22]. In addition, the co-channel interference among small cells is assumed to be a part of the thermal noise [23], since the cross-tier channel power gain is very weak because the penetration loss and the peak transmit power of each small cell user is usually limited because of regulation constraints [24].

Assume that each computation task can be described in four terms as $T_{u_s} = \{z_{u_s}, z'_{u_s}, c_{u_s}, R_{u_s}^{\mathrm{cp}}\}, \forall s, u$. For the task T_{u_s}, z_{u_o} and z'_{u_o}, respectively, represent the sizes of the contents before and after the computation. c_{u_s} denotes the computing ability required for accomplishing this task, which can be quantized by the amount of CPU cycles [6]. $R_{u_s}^{\mathrm{cp}}$ is the minimum computation rate required by user u_s.

Let e_n be the energy consumption for one CPU cycle at BS n. We denote $f_{u_s,n}$ as the computation capability of BS n assigned to user u_s, which is quantized by the total number of CPU cycles per second [6]. Practically, similar to mobile data usage service, $f_{u_s,n}$ can be determined based on the computation service contract subscribed to the mobile user by the telecom operator. Then the computation execution time of the task at BS n can be easily obtained as $t_{u_s,n} = \frac{c_{u_s}}{f_{u_s,n}}$. Therefore, the computation rate (i.e., the amount of bits computed during one second) of BS n to compute task T_{u_s} can be equivalent to

$$R_{u_s,n} = \frac{z_{u_s}}{t_{u_s,n}} = \frac{f_{u_s,n} z_{u_s}}{c_{u_s}}. \tag{4.4}$$

The total energy consumption used for computing task T_{u_s} at BE n can be calculated as

$$E_{u_s,n} = c_{u_s} e_n. \tag{4.5}$$

Each user has the requirement for the computation rate,

$$\sum_{n \in \mathcal{N}} a_{u_s,n} R_{u_s,n} \geq R_{u_s}^{\text{cp}}, \forall s \in \mathcal{S}, u_s \in \mathcal{U}_s. \tag{4.6}$$

Moreover, it should be noted that the computation ability at each BS is limited; thus

$$\sum_{s \in \mathcal{S}} \sum_{u_s \in \mathcal{U}_s} a_{u_s,n} \leq D_n, \forall n \in \mathcal{N}, \tag{4.7}$$

where D_n is the maximum amount of tasks simultaneously executed on the MEC server of BS n.

4.3.1.3 Caching model

For each BS, one can determine whether to cache the content sent by users before or after the computation, according to the popularity distribution of each content. The caching strategy can be controlled by two binary parameters $x_{u_s,n}^1$ and $x_{u_s,n}^2$. If BS n caches the content before the computation, $x_{u_s,n}^1 = 1$; otherwise $x_{u_s,n}^1 = 0$. If BS n caches the content after the computation, $x_{u_s,n}^2 = 1$; otherwise $x_{u_s,n}^2 = 0$. It should be noted that the storage of BS n may be limited. Thus, the cached content cannot be larger than the remaining space Z_n of BS n, which can be expressed as

$$\sum_{s \in \mathcal{S}} \sum_{u_s \in \mathcal{U}_s} a_{u_s,n}(x_{u_s,n}^1 z_{u_s} + x_{u_s,n}^2 z_{u_s}') \leq Z_n, \forall n \in \mathcal{N}. \tag{4.8}$$

In this chapter, it is assumed that the popularity distribution is represented by a vector $\boldsymbol{p} = [p_1, p_2, ..., p_F]$, where F types of contents with diverse popularity are distributed in the networks. That is, each content f is requested by each mobile user independently with the probability p_f. Generally, $\boldsymbol{p}$ is modeled as the Zipf distribution [25], which can be expressed as

$$p_f = \frac{1/f^\epsilon}{\sum\limits_{f=1}^{F} 1/f^\epsilon}, \forall f, \tag{4.9}$$

where the exponent ϵ is a positive value and can characterize content popularity. A higher value of ϵ corresponds to a higher level of content reuse. Typically, ϵ is between 0.5 and 1.5. For our business model, $p_{z_{u_s}}$ and $p_{z_{u_s}'}$ can be directly derived from p_f if the content sent by user u_s is known. Afterwards, the gains of the expected saved backhaul bandwidth through caching contents z_{u_s} and z_{u_s}' can be calculated as, respectively,

$$g_{z_{u_s}} = \frac{p_{z_{u_s}} z_{u_s}}{T_{z_{u_s}}}, \tag{4.10}$$

and

$$g_{z'_{u_s}} = \frac{p_{z'_{u_s}} z'_{u_s}}{T_{z'_{u_s}}}, \tag{4.11}$$

where $T_{z_{u_s}}$ and $T_{z'_{u_s}}$ are the time durations for downloading the required contents through backhaul.

In our business model, MVNO also needs to dynamically pay the usage of each BS to cache the content. It should be noted that the price for caching a content with a lower degree of popularity can be set as a higher value; thus some contents without popularity may not be cached at BSs. Actually, pricing for the caching system has been studied by many researchers. In this chapter, we assume that the price is already known, and we aim to present a caching strategy with a known price.

4.3.2 Problem formulation

In this subsection, an optimization problem is formulated to maximize the aggregate utility of the MVNO system. The optimization problem is mathematically modeled as

$$OP1: \max_{\substack{\{a_{u_s,n}, b_{u_s,n}, \\ x^1_{u_s,n}, x^2_{u_s,n}\}}} \sum_{s \in \mathcal{S}} \sum_{u_s \in \mathcal{U}_s} \sum_{n \in \mathcal{N}} U_{u_s,n}$$

$$s.t.: C1: \sum_{n \in \mathcal{N}} a_{u_s,n} = 1, \forall s \in \mathcal{S}, u_s \in \mathcal{U}_s$$

$$C2: \sum_{s \in \mathcal{S}} \sum_{u_s \in \mathcal{U}_s} a_{u_s,n} b_{u_s,n} \le B_n, \forall n \in \mathcal{N}$$

$$C3: \sum_{n \in \mathcal{N}} a_{u_s,n} b_{u_s,n} r_{u_s,n} \ge R^{\mathrm{cm}}_{u_s}, \forall s \in \mathcal{S}, u_s \in \mathcal{U}_s$$

$$C4: \sum_{n \in \mathcal{N}} a_{u_s,n} R_{u_s,n} \ge R^{\mathrm{cp}}_{u_s}, \forall s \in \mathcal{S}, u_s \in \mathcal{U}_s$$

$$C5: \sum_{s \in \mathcal{S}} \sum_{u_s \in \mathcal{U}_s} a_{u_s,n} \le D_n, \forall n \in \mathcal{N}$$

$$C6: \sum_{s \in \mathcal{S}} \sum_{u_s \in \mathcal{U}_s} a_{u_s,n} (x^1_{u_s,n} z_{u_s} + x^2_{u_s,n} z'_{u_s}) \le Z_n, \forall n \in \mathcal{N}$$

where $U_{u_s,n}$ is the potential utility of user u_s associating with BS n. Furthermore, according to constraint C1, each user can associate to only one BS. Constraint C2 reflects that the total amount of the allocated resource to all users associating with BS n cannot exceed the total radio resource of BS n. Constraints C3 and C4 guarantee the communition rate requirement and the computation rate requirement of each user, respectively. Constraint C5 restricts the computation ability of each BS. Constraint C6 ensures that the caching strategy is limited by the empty space of each BS.

Mathematically, the utility for the potential transmission between user u_s and BS n can be defined as

$$
\begin{aligned}
U_{u_s,n} = {} & a_{u_s,n}(\alpha_{u_s} b_{u_s,n} r_{u_s,n} - \beta_n b_{u_s,n}) \\
& + a_{u_s,n}(\phi_{u_s} R_{u_s,n} - \psi_n E_{u_s,n}) \\
& + a_{u_s,n} x^1_{u_s,n}(\gamma_n g_{z_{u_s}} - \Psi^n_{z_{u_s}} z_{u_s}) \\
& + a_{u_s,n} x^2_{u_s,n}(\gamma_n g_{z'_{u_s}} - \Psi^n_{z'_{u_s}} z'_{u_s}).
\end{aligned}
\tag{4.12}
$$

The utility function of $U_{u_s,n}$ consists of three parts, which can be further explained as follows:

- Communication Revenue: $\alpha_{u_s} b_{u_s,n} r_{u_s,n}$ denotes the income of MVNO from users to access the virtual networks, $\beta_n b_{u_s,n}$ denotes the expense of MVNO to pay for the usage of radio bandwidth, and $\alpha_{u_s} b_{u_s,n} r_{u_s,n} - \beta_n b_{u_s,n}$ is the communication revenue of MVNO if user u_s accesses the network n.

- Computation Revenue: $\phi_{u_s} R_{u_s,n}$ denotes the income of MVNO from users to access MEC servers for executing offloaded computation tasks, $\psi_n E_{u_s,n}$ denotes the expense of MVNO to pay for the usage of MEC servers, and $\phi_{u_s} R_{u_s,n} - \psi_n E_{u_s,n}$ is the computation revenue of MVNO if user u_s chooses to execute the computation task at BS n.

- Caching Revenue: $\gamma_n g_{z_{u_s}}$ and $\gamma_n g_{z'_{u_s}}$ are the saving expenses of MVNO, which are achieved on the saved backhaul bandwidth from caching the contents z_{u_s} and z'_{u_s}, and $\Psi^n_{z_{u_s}} z_{u_s}$ and $\Psi^n_{z'_{u_s}} z'_{u_s}$ are the expenses of MVNO to store the contents z_{u_s} and z'_{u_s} at BSs. Thus $\gamma_n g_{z_{u_s}} - \Psi^n_{z_{u_s}} z_{u_s}$ and $\gamma_n g_{z'_{u_s}} - \Psi^n_{z'_{u_s}} z'_{u_s}$ are the caching revenue if BS n chooses to cache the contents z_{u_s} and z'_{u_s}, respectively.

4.3.3 Problem reformulation

It is obvious that the formulated mixed discrete and non-convex optimization problem is an NP-hard problem; thus it is very challenging to find its global optimal solution. In this subsection, we reformulate the problem OP1 to make it tractable. A relaxation of the binary conditions of $a_{u_s,n}$, $x^1_{u_s,n}$, and $x^2_{u_s,n}$ constitutes the first step to solve the problem OP1, where $a_{u_s,n}$, $x^1_{u_s,n}$, and $x^2_{u_s,n}$ are relaxed to be real value variables as $0 \leq a_{u_s,n} \leq 1$, $0 \leq x^1_{u_s,n} \leq 1$ and $0 \leq x^2_{u_s,n} \leq 1$. The relaxed $a_{u_s,n}$ is sensible and meaningful as a time sharing factor representing the ratio of the time for user u_s to associate with BS n in order to offload and compute the offloading task. The relaxed $x^1_{u_s,n}$ and $x^2_{u_s,n}$ can also be interpreted as the time fractions for sharing one unit cache of BS n.

However, even after the relaxation of the variables, the problem is still non-convex due to the multiplication of the variables. Thus, a second step is necessary for further simplifying the problem to make it tractable and solvable. The following is a given proposition of the equivalent problem of OP1.

Proposition 4.1: If we define $\widetilde{x}^1_{u_s,n} = a_{u_s,n}x^1_{u_s,n}$, $\widetilde{x}^2_{u_s,n} = a_{u_s,n}x^2_{u_s,n}$, and $\widetilde{b}_{u_s,n} = a_{u_s,n}b_{u_s,n}$, there exists an equivalent formulation of problem OP1 as follows:

$$OP2: \max_{\substack{\{a_{u_s,n}, \widetilde{b}_{u_s,n}, \\ \widetilde{x}^1_{u_s,n}, \widetilde{x}^2_{u_s,n}\}}} \sum_{s\in\mathcal{S}} \sum_{u_s\in\mathcal{U}_s} \sum_{n\in\mathcal{N}} \widetilde{U}_{u_s,n}$$

$$s.t.: C1: \sum_{n\in\mathcal{N}} a_{u_s,n} = 1, \forall s\in\mathcal{S}, u_s\in\mathcal{U}_s$$

$$C2: \sum_{s\in\mathcal{S}} \sum_{u_s\in\mathcal{U}_s} \widetilde{b}_{u_s,n} \leq B_n, \forall n\in\mathcal{N}$$

$$C3: \sum_{n\in\mathcal{N}} \widetilde{b}_{u_s,n} r_{u_s,n} \geq R^{\mathrm{cm}}_{u_s}, \forall s\in\mathcal{S}, u_s\in\mathcal{U}_s$$

$$C4: \sum_{n\in\mathcal{N}} a_{u_s,n} R_{u_s,n} \geq R^{\mathrm{cp}}_{u_s}, \forall s\in\mathcal{S}, u_s\in\mathcal{U}_s$$

$$C5: \sum_{s\in\mathcal{S}} \sum_{u_s\in\mathcal{U}_s} a_{u_s,n} \leq D_n, \forall n\in\mathcal{N}$$

$$C6: \sum_{s\in\mathcal{S}} \sum_{u_s\in\mathcal{U}_s} (\widetilde{x}^1_{u_s,n} z_{u_s} + \widetilde{x}^2_{u_s,n} z'_{u_s}) \leq Z_n, \forall n\in\mathcal{N}$$

where $\widetilde{U}_{u_s,n}$ is expressed as

$$\widetilde{U}_{u_s,n} = \widetilde{b}_{u_s,n}(\alpha_{u_s} r_{u_s,n} - \beta_n) + a_{u_s,n}(\phi_{u_s} R_{u_s,n} - \psi_n E_{u_s,n}) \\ + \widetilde{x}^1_{u_s,n}(\gamma_n g_{z_{u_s}} - \Psi^n_{z_{u_s}} z_{u_s}) + \widetilde{x}^2_{u_s,n}(\gamma_n g_{z'_{u_s}} - \Psi^n_{z'_{u_s}} z'_{u_s}). \tag{4.13}$$

The relaxed problem OP1 can be directly recovered through substituting the variables $\widetilde{x}^1_{u_s,n} = a_{u_s,n}x^1_{u_s,n}$, $\widetilde{x}^2_{u_s,n} = a_{u_s,n}x^2_{u_s,n}$, and $\widetilde{b}_{u_s,n} = a_{u_s,n}b_{u_s,n}$ into problem OP2. If $a_{u_s,n} = 0$, $b_{u_s,n} = 0$ certainly holds due to optimality. Obviously, there is no need for BS n to allocate any resource to a user when the user does not associate with BS n. Afterwards, the mapping between $\{a_{u_s,n}, b_{u_s,n}, x^1_{u_s,n}, x^2_{u_s,n}\}$ and $\{a_{u_s,n}, \widetilde{b}_{u_s,n}, \widetilde{x}^1_{u_s,n}, \widetilde{x}^2_{u_s,n}\}$ can be easily obtained as

$$b_{u_s,n} = \begin{cases} \dfrac{\widetilde{b}_{u_s,n}}{a_{u_s,n}}, & a_{u_s,n} > 0 \\ 0, & \text{otherwise} \end{cases} \tag{4.14}$$

$$x^1_{u_s,n} = \begin{cases} \dfrac{\widetilde{x}^1_{u_s,n}}{a_{u_s,n}}, & a_{u_s,n} > 0 \\ 0, & \text{otherwise} \end{cases} \tag{4.15}$$

and

$$x^2_{u_s,n} = \begin{cases} \dfrac{\widetilde{x}^2_{u_s,n}}{a_{u_s,n}}, & a_{u_s,n} > 0 \\ 0, & \text{otherwise} \end{cases} \tag{4.16}$$

Now problem OP2 is transformed as a convex problem. A lot of methods, such as the interior point method and dual decomposition, can be used for solving such a convex problem. However, when a centralized algorithm is used to solve the problem, the signaling overhead will be prohibitively large, especially when the amounts of BSs and users are excessive, because finding the optimal solution requires all the CSI and content distribution information. Therefore, a distributed optimization algorithm executed on each BS is necessary to be designed for practical implementation. However, because of the constraints $C1, C3$, and $C4$, problem OP2 is not separable to be executed on each BS. Thus, coupling has to be decoupled appropriately, which will be discussed in the following section. To lighten the notation, from now on, u is used to denote each user instead of u_s.

4.4 Resource allocation via ADMM

In this section, each step of the decentralized algorithm is described in detail.

4.4.1 Decoupling of association indicators

In order to decouple the coupling variables, the local copies of $\{a_{u,n}\}$ and $\{\widetilde{b}_{u,n}\}$ at BS n is introduced as $\{\widehat{a}_{u,k}^n\}$ and $\{\widehat{b}_{u,k}^n\}$, respectively. With the local vectors $\{\widehat{a}_{u,k}^n\}$ and $\{\widehat{b}_{u,k}^n\}$, a feasible local variable set for each BS n can be defined as

$$
\mathcal{X}_n = \left\{
\begin{array}{c}
\{\widehat{a}_{u,k}^n\} \\
\{\widehat{b}_{u,k}^n\}
\end{array}
\left|
\begin{array}{c}
\sum_{k\in\mathcal{N}} \widehat{a}_{u,k}^n = 1, \forall u \\
\sum_{u\in\mathcal{U}} \widehat{b}_{u,k}^n \leq B_k, \forall k \\
\sum_{k\in\mathcal{N}} \widehat{b}_{u,k}^n r_{u,k} \geq R_u^{\mathrm{cm}}, \forall u \\
\sum_{k\in\mathcal{N}} \widehat{a}_{u,k}^n R_{u,k} \geq R_u^{\mathrm{cp}}, \forall u \\
\sum_{u\in\mathcal{U}} \widehat{a}_{u,k}^n \leq D_k, \forall k \\
\sum_{u\in\mathcal{U}} (\widetilde{x}_{u,k}^1 z_u + \widetilde{x}_{u,k}^2 z'_u) \leq Z_k, \forall k
\end{array}
\right.
\right\},
\tag{4.17}
$$

and an associated local utility function can be expressed as

$$
\dagger_n = \left\{
\begin{array}{ll}
-\sum_{u\in\mathcal{U}} \widehat{U}_{u,n}, (\{\widehat{a}_{u,k}^n\}, \{\widetilde{x}_{u,k}^1\}, \{\widetilde{x}_{u,k}^2\}, \{\widehat{b}_{u,k}^n\}) \in \mathcal{X}_n \\
0, \qquad \text{Otherwise}
\end{array}
\right.
\tag{4.18}
$$

where

$$
\widehat{U}_{u,n} = \widehat{b}_{u,k}^n(\alpha_u r_{u,k} - \beta_k) + \widehat{a}_{u,k}^n(\phi_u R_{u,k} - \psi_k E_{u,k}) \\
+ \widetilde{x}_{u,k}^1(\gamma_k g_{z_u} - \Psi_{z_u}^k z_u) + \widetilde{x}_{u,k}^2(\gamma_k g_{z'_u} - \Psi_{z'_u}^k z'_u).
\tag{4.19}
$$

With this notation, the global consensus problem of the problem OP2 can be shown as follows:

$$OP3 : \min \mathcal{Y}(\{\widehat{a}_{u,k}^n\}, \{\widetilde{x}_{u,k}^1\}, \{\widetilde{x}_{u,k}^2\}, \{\widehat{b}_{u,k}^n\}) =$$

$$\sum_{n \in \mathcal{N}} \dagger_n(\{\widehat{a}_{u,k}^n\}, \{\widetilde{x}_{u,k}^1\}, \{\widetilde{x}_{u,k}^2\}, \{\widehat{b}_{u,k}^n\})$$

$$s.t. : \{\widehat{a}_{u,k}^n\} = \{a_{u,k}\}, \{\widehat{b}_{u,k}^n\} = \{\widetilde{b}_{u,k}\}, \forall n, u, k$$

Obviously, now the objective function is separable across each BS, but the global association variables are still involved in the constraints. Therefore, an ADMM-based algorithm is introduced for solving the optimization problem OP3 in a distributed manner, which is shown in the following subsection.

4.4.2 Problem solving via ADMM

The proposed algorithm for a virtual resource allocation scheme via ADMM is described in this subsection. First, the initial step of ADMM to solve the problem OP3 is the formulation of an augmented Lagrangian $\mathcal{L}_\rho(\{\widehat{a}, \widetilde{x}^1, \widetilde{x}^2, \widehat{b}\}, \{a, \widetilde{b}\}, \{\mu, \nu\})$ with corresponding global consensus constraints. Here, $\widehat{a} = \{\widehat{a}_{u,k}^n\}$, $\widetilde{x}^1 = \{\widetilde{x}_{u,k}^1\}$, $\widetilde{x}^2 = \{\widetilde{x}_{u,n}^2\}$, $\widehat{b} = \{\widehat{b}_{u,k}^n\}$, $a = \{a_{u,k}\}$, and $\widetilde{b} = \{\widetilde{b}_{u,k}\}$. The augmented Lagrangian can be derived as [14]

$$\mathcal{L}_\rho(\{\widehat{a}, \widetilde{x}^1, \widetilde{x}^2, \widehat{b}\}, \{a, \widetilde{b}\}, \{\mu, \nu\}) = \sum_{n \in \mathcal{N}} \dagger_n(\widehat{a}^n, \widetilde{x}^1, \widetilde{x}^2, \widehat{b}^n) +$$

$$\sum_{n \in \mathcal{N}} \sum_{\substack{u \in \mathcal{U} \\ k \in \mathcal{N}}} \mu_{u,k}^n(\widehat{a}_{u,k}^n - a_{u,k}) + \frac{\rho}{2} \sum_{n \in \mathcal{N}} \sum_{\substack{u \in \mathcal{U} \\ k \in \mathcal{N}}} (\widehat{a}_{u,k}^n - a_{u,k})^2 + \qquad (4.20)$$

$$\sum_{n \in \mathcal{N}} \sum_{\substack{u \in \mathcal{U} \\ k \in \mathcal{N}}} \nu_{u,k}^n(\widehat{b}_{u,k}^n - \widetilde{b}_{u,k}) + \frac{\rho}{2} \sum_{n \in \mathcal{N}} \sum_{\substack{u \in \mathcal{U} \\ k \in \mathcal{N}}} (\widehat{b}_{u,k}^n - \widetilde{b}_{u,k})^2,$$

where ρ is the penalty parameter, and $\mu = \{\mu_{u,k}^n\}$ and $\nu = \{\nu_{u,k}^n\}$ are the dual variables. The convergence of the ADMM-based algorithm is limited by the penalty parameter ρ. A larger value of ρ will make the primal dual quickly converge to zero, but it will also result in an increased dual residual. Thus, a proper value of ρ is of enormous importance to control the process of the ADMM-based algorithm.

Compared to the standard Lagrangian, the performance of the iterative method can be improved by adding the quadratic penalty term in objective function [26]. Now, the solution to the minimization of (4.20) is equivalent to that of the original problem OP3, because the penalty term added to the objective function is actually equal to zero for any feasible solution [14].

According to the iteration of AMDD with consensus constraints, the process for solving the problem OP3 consists of the following steps:

Step 1: $\{\widehat{a}^n, \widetilde{x}^1, \widetilde{x}^2, \widehat{b}^n\}$ − *update:* In this step, which can be regarded as local variables updating, the problem OP3 can be completely decoupled into

N specific subproblems, and each of the subproblems can be solved locally and separately at BSs according to

$$\{\widehat{\boldsymbol{a}}^n, \widetilde{\boldsymbol{x}}^1, \widetilde{\boldsymbol{x}}^2, \widehat{\boldsymbol{b}}^n\}_{n\in\mathcal{N}}^{[i+1]} := \arg\min\{\dagger_n(\widehat{\boldsymbol{a}}^n, \widetilde{\boldsymbol{x}}^1, \widetilde{\boldsymbol{x}}^2, \widehat{\boldsymbol{b}}^n)$$

$$+ \sum_{\substack{u\in\mathcal{U}\\k\in\mathcal{N}}} \mu_{u,k}^{n[i]}(\widehat{a}_{u,k}^n - a_{u,k}^{[i]}) + \frac{\rho}{2}\sum_{\substack{u\in\mathcal{U}\\k\in\mathcal{N}}} (\widehat{a}_{u,k}^n - a_{u,k}^{[i]})^2 \tag{4.21}$$

$$+ \sum_{\substack{u\in\mathcal{U}\\k\in\mathcal{N}}} \nu_{u,k}^{n[i]}(\widehat{b}_{u,k}^n - \widetilde{b}_{u,k}^{[i]}) + \frac{\rho}{2}\sum_{\substack{u\in\mathcal{U}\\k\in\mathcal{N}}} (\widehat{b}_{u,k}^n - \widetilde{b}_{u,k}^{[i]})^2\}.$$

After eliminating the constant term, each BS n solves the following optimization problem at iteration $[i]$:

$$OP4: \min \mathcal{L}_\rho^n = -\sum_{u\in\mathcal{U}} \widehat{U}_{u,n} +$$

$$\sum_{\substack{u\in\mathcal{U}\\k\in\mathcal{N}}} [\mu_{u,k}^{n[i]}\widehat{a}_{u,k}^n + \frac{\rho}{2}(\widehat{a}_{u,k}^n - a_{u,k}^{[i]})^2] +$$

$$\sum_{\substack{u\in\mathcal{U}\\k\in\mathcal{N}}} [\nu_{u,k}^{n[i]}\widehat{b}_{u,k}^n + \frac{\rho}{2}(\widehat{b}_{u,k}^n - \widetilde{b}_{u,k}^{[i]})^2],$$

$$s.t.: (\widehat{\boldsymbol{a}}^n, \widetilde{\boldsymbol{x}}^1, \widetilde{\boldsymbol{x}}^2, \widehat{\boldsymbol{b}}^n) \in \mathcal{X}_n$$

In this chapter, the primal dual interior-point method, which is able to provide an efficient way for solving convex problems [27], is used to find the optimal solution of the problem. Due to limited space, the details of the procedure are omitted here.

Step 2: $\{\boldsymbol{a}, \widetilde{\boldsymbol{b}}\}-update:$ The second step is global variables updating, where $\boldsymbol{a}$ and $\widetilde{\boldsymbol{b}}$ can be updated according to

$$\boldsymbol{a}^{[i+1]} := \arg\min \sum_{n\in\mathcal{N}} \sum_{\substack{u\in\mathcal{U}\\k\in\mathcal{N}}} \mu_{u,k}^{n[i]}(\widehat{a}_{u,k}^{n[i+1]} - a_{u,k})$$

$$+ \frac{\rho}{2}\sum_{n\in\mathcal{N}} \sum_{\substack{u\in\mathcal{U}\\k\in\mathcal{N}}} (\widehat{a}_{u,k}^{n[i+1]} - a_{u,k})^2,$$

$$\widetilde{\boldsymbol{b}}^{[i+1]} := \arg\min \sum_{n\in\mathcal{N}} \sum_{\substack{u\in\mathcal{U}\\k\in\mathcal{N}}} \nu_{u,k}^{n[i]}(\widehat{b}_{u,k}^{n[i+1]} - \widetilde{b}_{u,k}) \tag{4.22}$$

$$+ \frac{\rho}{2}\sum_{n\in\mathcal{N}} \sum_{\substack{u\in\mathcal{U}\\k\in\mathcal{N}}} (\widehat{b}_{u,k}^{n[i+1]} - \widetilde{b}_{u,k})^2.$$

Because we have added the quadratic regularization term to the augmented Lagrangian (4.20), the unconstrained problems (4.22) are strictly convex with respect to $\boldsymbol{a}$ and $\widetilde{\boldsymbol{b}}$. Through setting the gradients to zero, (4.22) can be simplified as

$$a_{u,k}^{[i+1]} = \frac{1}{N}\sum_{n\in\mathcal{N}} [\widehat{a}_{u,k}^{n[i+1]} + \frac{1}{\rho}\mu_{u,k}^{n[i]}], \forall u, k,$$

$$\widetilde{b}_{u,k}^{[i+1]} = \frac{1}{N}\sum_{n\in\mathcal{N}} [\widehat{b}_{u,k}^{n[i+1]} + \frac{1}{\rho}\nu_{u,k}^{n[i]}], \forall u, k. \tag{4.23}$$

To initialize the dual variables to zeros, we have $\sum_{n\in\mathcal{N}} \mu_{u,k}^{n[i]} = 0$ and $\sum_{n\in\mathcal{N}} \nu_{u,k}^{n[i]} = 0$, $\forall u, k$, at each iteration $[i]$ [26]. Afterwards, (4.23) is reduced to

$$
\begin{aligned}
a_{u,k}^{[i+1]} &= \frac{1}{N} \sum_{n\in\mathcal{N}} \widehat{a}_{u,k}^{n[i+1]}, \forall u, k, \\
\widetilde{b}_{u,k}^{[i+1]} &= \frac{1}{N} \sum_{n\in\mathcal{N}} \widehat{b}_{u,k}^{n[i+1]}, \forall u, k.
\end{aligned}
\tag{4.24}
$$

This step can be optimized by a central controller of MVNO, and it can be viewed as the procedure of gathering all the updated local copies and averaging them out. It should be noted that this step does not involve dual variables, which will result in a significantly decreased signaling overhead while exchanging information [12].

Step 3: $\{\boldsymbol{\mu}, \boldsymbol{\nu}\}$ − *update:* This step shows the updating of dual variables, which can be represented as

$$
\begin{aligned}
\boldsymbol{\mu}^{n[i+1]} &:= \boldsymbol{\mu}^{n[i]} + \rho(\widehat{\boldsymbol{a}}^{n[i+1]} - \boldsymbol{a}^{[i+1]}), \\
\boldsymbol{\nu}^{n[i+1]} &:= \boldsymbol{\nu}^{n[i]} + \rho(\widehat{\boldsymbol{b}}^{n[i+1]} - \widetilde{\boldsymbol{b}}^{[i+1]}).
\end{aligned}
\tag{4.25}
$$

Here, the augmented Lagrangian parameter ρ is used as a step size to update the dual variables.

Step 4: Algorithm Stopping Criterion: The rational stopping criterion proposed in [14] is introduced in this step, which is given as

$$
\begin{aligned}
||r_p^{[i+1]}||_2 &\leq \upsilon_{pri}, \\
||r_d^{[i+1]}||_2 &\leq \upsilon_{dual},
\end{aligned}
\tag{4.26}
$$

where $\upsilon_{pri} > 0$ and $\upsilon_{dual} > 0$, called the feasibility tolerances of the primal feasibility and dual feasibility conditions, respectively. This stopping criterion guarantees that the primal residual $r_p^{[i+1]}$ and the dual residual $r_d^{[i+1]}$ of the final solution are small enough.

It has been demonstrated in [14] that the final solution satisfies the second dual feasibility condition through updating variables $\boldsymbol{\mu}$ and $\boldsymbol{\nu}$ via Step 3, where the second dual feasibility condition can be obtained by taking gradients with respect to global variables. Actually, the primal feasibility and the first dual feasibility, which can be obtained by taking gradients with respect to local variables, do not hold. However, the primal residual and first dual residual are able to converge to zero, and this means that the primal feasibility and the first dual feasibility are achieved when $i \to \infty$.

According to (4.26), the residual for the primal feasibility condition of BS n at iteration $[i]$ should be small enough so that

$$
\begin{aligned}
||\widehat{\boldsymbol{a}}^{n[i+1]} - \boldsymbol{a}^{[i+1]}||_2 &\leq \upsilon_{pri}, \\
||\widehat{\boldsymbol{b}}^{n[i+1]} - \widetilde{\boldsymbol{b}}^{[i+1]}||_2 &\leq \upsilon_{pri}.
\end{aligned}
\tag{4.27}
$$

Moreover, the residual for the first dual feasibility condition at iteration $[i+1]$

should be small enough so that

$$
\begin{aligned}
||\boldsymbol{a}^{[i+1]} - \boldsymbol{a}^{[i]}||_2 &\leq v_{dual}, \\
||\widetilde{\boldsymbol{b}}^{[i+1]} - \widetilde{\boldsymbol{b}}^{[i]}||_2 &\leq v_{dual}.
\end{aligned}
\tag{4.28}
$$

Step 5: $\{\boldsymbol{a}, \widetilde{\boldsymbol{x}}^1, \widetilde{\boldsymbol{x}}^2\} - recovery$: This step is the recovery of $\boldsymbol{a}, \widetilde{\boldsymbol{x}}^1$, and $\widetilde{\boldsymbol{x}}^2$, since we have relaxed the variables between zero and one instead of binary variables in Subsection 4.3.3. In the proposed algorithm, we recover $\boldsymbol{a}, \widetilde{\boldsymbol{x}}^1$, and $\widetilde{\boldsymbol{x}}^2$ to binary after obtaining the optimum solution. The binary recovery can be viewed as computing the marginal benefit for each user u [28]. Then, the variables can be recovered as

$$
a_{u^*,k} = \begin{cases} 1, & \text{If } u^* = \arg\max\{Q_{u,k} > 0, \forall u\} \\ 0, & \text{otherwise} \end{cases}
\tag{4.29}
$$

Here, $Q_{u,k}$ is the first partial derivation of $U_{u,k}$ with respect to $a_{u,k}$. $x^1_{u,k} = 1$ and $x^2_{u,k} = 1$ make sense only when $a_{u,k} = 1$. Therefore, if finally we find that $a_{u,k} = 1$, we set $x^1_{u,k} = 1$ and $x^2_{u,k} = 1$ unless $x^1_{u,k} = 0$ and $x^2_{u,k} = 0$ from the ADMM process.

4.4.3 Algorithm analysis: computational complexity

In this subsection, we compare the computational complexity of the proposed distributed algorithm with that of the centralized algorithm. For the centralized algorithm of the primal dual interior-point method, the computational complexity is typically $O(((N + 1)U)^k)$. However, for the proposed ADMM-based distributed algorithm, each BS only needs to solve its own optimization subproblem, and thus the computational complexity at each BS is $O(U^k)$, where $k = 1$ represents a linear algorithm and $k > 1$ represents a polynomial time algorithm. After solving the corresponding subproblems, the central controller of MVNO needs to collect the local solutions and calculate the global solution with the computational complexity $O((N + 1)U)$, and it also needs to update the dual variables with the complexity $O((N + 1)U)$. Assume that the iteration number is I, and thus the total computational complexity of the distributed algorithm is reduced to $I(O((N + 1) \cdot U^k) + 2(N + 1)U) = I(O((N + 1) \cdot U^k))$.

4.5 Simulation results and discussion

In this section, simulations are given to investigate different aspects of the virtual system and to evaluate the performance of the proposed algorithm.

Table 4.2: Parameter Settings

Parameter	Value
Bandwidth allocated to MBS (B_0)	10MHz
Bandwidth allocated to SBS n (B_n)	5MHz
Communication rate requirement of each user ($R_{u_s}^{\mathrm{cm}}$)	10Mbps
Computation rate requirement of each user ($R_{u_s}^{\mathrm{cp}}$)	1Mbps
Energy consumption for one CPU cycle at BS n (e_n)	1W/GHz
Total workload of MBS (D_0)	100
Total workload of SBS (D_n)	50
Spectrum usage cost of MBS (β_0)	3units/KHz
Spectrum usage cost of SBS n (β_n)	1units/KHz
Computation fee of MBS (ψ_0)	$80 * 10^{-6}$units/J
Computation fee of BS n (ψ_n)	$40 * 10^{-6}$units/J
Storage fee of MBS ($\Psi_{z_{u_s}}^0$ or $\Psi_{z'_{u_s}}^0$)	20units/Mb
Storage fee of BS n ($\Psi_{z_{u_s}}^n$ or $\Psi_{z'_{u_s}}^n$)	10units/Mb
PSD of AWGN (σ)	-174dBm
Transmit power of each user (q_{u_s})	27dBm

4.5.1 Parameter settings

We assume that there are 2 SPs, 9 BSs, 1 MVNO, and 50 users. The wireless channels between mobile users and BSs suffer from Rayleigh fading. All the channel coefficients are distributed as $\mathcal{CN}(0, \frac{1}{(1+d)^\alpha})$ with a path loss exponent $\alpha = 4$, where d is the distance between each mobile user and a BS. The values of the rest of parameters are summarized in Table 4.2. Similar to the previous works [12, 13, 25, 29], the well-known Monte Carlo simulation is used for evaluating the performance of the proposed scheme. In this chapter, MATLAB 7.9.0 (R2009b) is adopted to present the performance evaluation, and a WIN7 desktop computer with quad-core CPU (Intel Q8400) and 4-GB RAM is used to run the simulations through the Monte Carlo method. We assume the position of MBS is fixed, and the positions of all SBSs are uniformly distributed within the covered area of the MBS. Based on the Monte Carlo simulation, the locations of users are changed in each simulation loop. After several loops, the average value is calculated to reduce randomness effects.

4.5.2 Alternative schemes

For performance comparison, two other schemes are also evaluated, the details of which are as follows:

■ Centralized Scheme with (w.) Caching: collects content distribution information and CSI from all users and executes a virtual resource allocation strategy in a centralized manner.

■ Distributed Scheme without (w.o.) Caching: each BS only needs to solve its own problem without exchange of CSI but cannot support content caching in this compared scheme.

4.5.3 Performance evaluation

Figures 4.3 and 4.4 show the convergence performance of the proposed scheme and the alternative schemes under different numbers of users and BSs. It is obvious that all schemes are able to converge to a stable solution rapidly, and the proposed scheme with different values of ρ can eventually converge to the same value of the total utility of MVNO. However, a higher value of ρ will result in a higher rate of convergence. Thus, in the following performance evaluation, we set $\rho = 200$. Furthermore, we can observe that the proposed scheme performs better than the distributed scheme without the caching function. Although there is a performance gap from the centralized scheme, the advantage of the proposed scheme is the reduced signal overhead for the exchange of content distribution information and the CSI.

Figures 4.5 to 4.7 illustrate the total utility, computation revenue, and caching revenue of different schemes with respect to the different values of total number of users. As the number of users increases, the total utilities, computation revenues, and caching revenues of the proposed scheme and the centralized scheme continue to grow. The main reason for the performance of the distributed scheme without the caching function being worse than the proposed scheme is that popular contents cannot be cached at BSs so that there is no caching revenue when some users call for previous contents. On the other hand, in the proposed scheme, if the contents after computation required by users have already been cached at the associated BSs, the BSs do not need to compute the offloading contents, which will certainly contribute to increasing computation revenue. Furthermore, the caching revenue of the proposed scheme is higher than that of the centralized scheme. The computation revenues are almost the same among these three schemes, because here we assume that each BS does not cache any content initially.

Figures 4.8 and 4.9 show the performance of the proposed scheme under different parameters of computation and caching functions. From Figure 4.8, it can be observed that the total utility of MVNO is decreased exponentially with the computation ability required for accomplishing each task, since a higher value of computation ability means a larger number of required

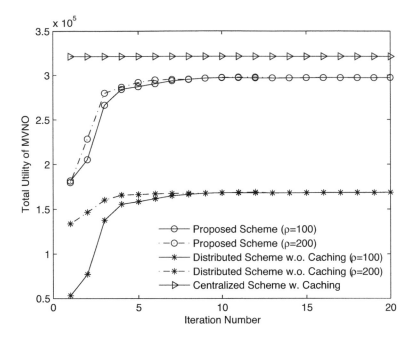

Figure 4.3: Convergence performance. The total number of users is $U = 30$. The total number of SBSs is $N = 4$. The sizes of each content, z_{u_s} and z'_{u_s}, are randomly distributed within 5Mb to 125Mb. The computing ability required for accomplishing each task, c_{u_s}, is distributed within 1Gegacycle to 13Gegacycles. The computation capability of MBS and each SBS are $f_{u_s,0} = 10$GHz and $f_{u_s,n} = 5$GHz, respectively, and the remaining spaces of MBS and each SBS are $Z_0 = 500$Mb and $Z_n = 250$Mb. The access fee from each user is $\alpha_{u_s} = 100$units/Mbps, the backhaul cost of MBS is $\gamma_0 = 12$units/Mbps, the backhaul cost of SBS n is $\gamma_n = 10$units/Mbps, and the computation fee from each user is $\phi_{u_s} = 100$units/Mbps.

CPU cycles, which will induce a higher consumed computation energy and a lower gain of computation rate. In addition, as the computation capability decreases, the computation rate is reduced, and thus the total utility of MVNO is decreased. From Figure 4.9, it is obvious that the total utility of MVNO is increased linearly with the size of each content. This is because caching a larger size of content will tremendously reduce backhaul bandwidth consumption, and content that needs to be computed with a larger size will cause a higher computation rate. Moreover, the total utility of MVNO is limited by the remaining spaces of each BS, since some popular content may not be able to be cached at the BSs with fewer caching spaces.

Figures 4.10 to 4.12 illustrate the performance of the proposed scheme and the two compared schemes under different values of prices. The total utilities

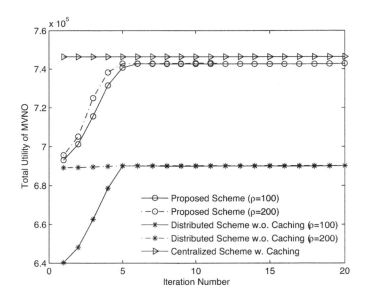

Figure 4.4: Convergence performance. $U = 40$, $N = 8$, $z_{u_s}, z'_{u_s} \in [5, 125]$Mb, $c_{u_s} \in [1, 13]$Gegacycles, $f_{u_s,0} = 10$GHz, $f_{u_s,n} = 5$GHz, $Z_0 = 500$Mb, $Z_n = 250$Mb, $\alpha_{u_s} = 100$units/Mbps, $\gamma_0 = 12$units/Mbps, $\gamma_n = 10$units/Mbps, $\phi_{u_s} = 100$units/Mbps.

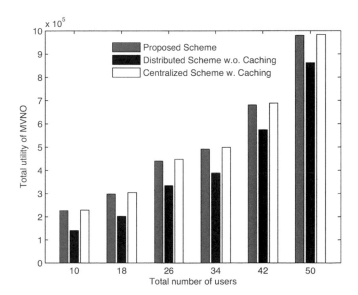

Figure 4.5: Total utility of MVNO with different numbers of users. $N = 8$, $z_{u_s}, z'_{u_s} \in [5, 125]$Mb, $c_{u_s} \in [1, 13]$Gegacycles, $f_{u_s,0} = 10$GHz, $f_{u_s,n} = 5$GHz, $Z_0 = 500$Mb, $Z_n = 250$Mb, $\alpha_{u_s} = 100$units/Mbps, $\gamma_0 = 12$units/Mbps, $\gamma_n = 10$units/Mbps, $\phi_{u_s} = 100$units/Mbps.

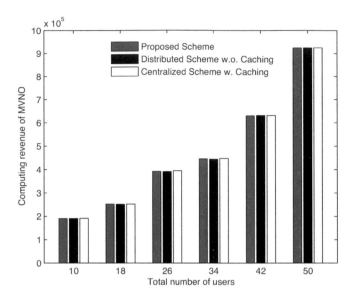

Figure 4.6: Comparison of computation revenue under different numbers of users. $N = 8$, $z_{u_s}, z'_{u_s} \in [5, 125]$Mb, $c_{u_s} \in [1, 13]$Gegacycles, $f_{u_s,0} = 10$GHz, $f_{u_s,n} = 5$GHz, $Z_0 = 500$Mb, $Z_n = 250$Mb, $\alpha_{u_s} = 100$units/Mbps, $\gamma_0 = 12$units/Mbps, $\gamma_n = 10$units/Mbps, $\phi_{u_s} = 100$units/Mbps.

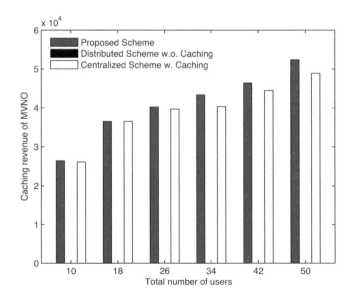

Figure 4.7: Comparison of caching revenue under different numbers of users. $N = 8$, $z_{u_s}, z'_{u_s} \in [5, 125]$Mb, $c_{u_s} \in [1, 13]$Gegacycles, $f_{u_s,0} = 10$GHz, $f_{u_s,n} = 5$GHz, $Z_0 = 500$Mb, $Z_n = 250$Mb, $\alpha_{u_s} = 100$units/Mbps, $\gamma_0 = 12$units/Mbps, $\gamma_n = 10$units/Mbps, $\phi_{u_s} = 100$units/Mbps.

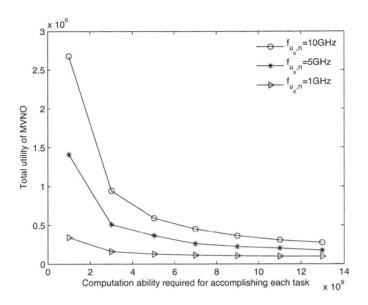

Figure 4.8: Performance evaluation under different parameter setting of computation function. $U = 40$, $N = 8$, $z_{u_s}, z'_{u_s} \in [5, 125]$Mb, $Z_0 = 500$Mb, $Z_n = 250$Mb, $\alpha_{u_s} = 100$units/Mbps, $\gamma_0 = 12$units/Mbps, $\gamma_n = 10$units/Mbps, $\phi_{u_s} = 100$units/Mbps.

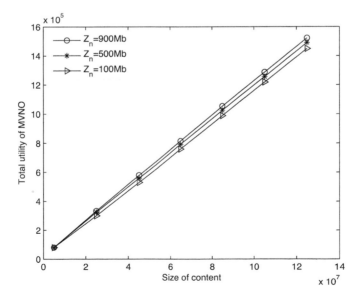

Figure 4.9: Performance evaluation under different parameter setting of caching function. $U = 40$, $N = 8$, $c_{u_s} \in [1, 13]$Gegacycles, $f_{u_s,0} = 10$GHz, $f_{u_s,n} = 5$GHz, $\alpha_{u_s} = 100$units/Mbps, $\gamma_0 = 12$units/Mbps, $\gamma_n = 10$units/Mbps, $\phi_{u_s} = 100$units/Mbps.

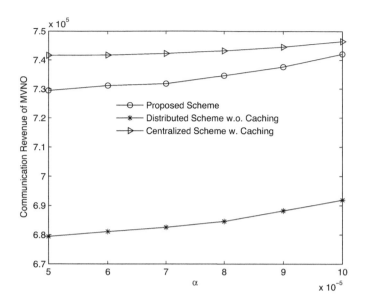

Figure 4.10: Total utility of MVNO with different values of α. $U = 40$, $N = 8$, $z_{u_s}, z'_{u_s} \in [5, 125]$Mb, $c_{u_s} \in [1, 13]$Gegacycles, $f_{u_s,0} = 10$GHz, $f_{u_s,n} = 5$GHz, $Z_0 = 500$Mb, $Z_n = 250$Mb, $\gamma_0 = 12$units/Mbps, $\gamma_n = 10$units/Mbps, $\phi_{u_s} = 100$units/Mbps.

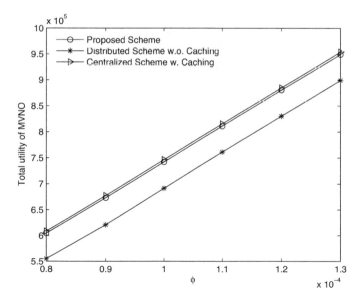

Figure 4.11: Total utility of MVNO with different values of ϕ. $U = 40$, $N = 8$, $z_{u_s}, z'_{u_s} \in [5, 125]$Mb, $c_{u_s} \in [1, 13]$Gegacycles, $f_{u_s,0} = 10$GHz, $f_{u_s,n} = 5$GHz, $Z_0 = 500$Mb, $Z_n = 250$Mb, $\alpha_{u_s} = 100$units/Mbps, $\gamma_0 = 12$units/Mbps, $\gamma_n = 10$units/Mbps.

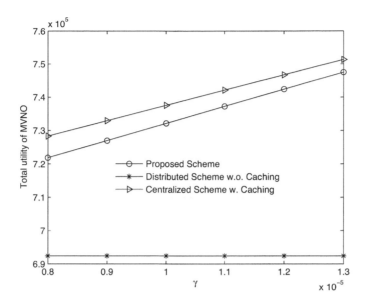

Figure 4.12: Total utility of MVNO with different values of γ. $U = 40$, $N = 8$, $z_{u_s}, z'_{u_s} \in [5, 125]$Mb, $c_{u_s} \in [1, 13]$Gegacycles, $f_{u_s,0} = 10$GHz, $f_{u_s,n} = 5$GHz, $Z_0 = 500$Mb, $Z_n = 250$Mb, $\alpha_{u_s} = 100$units/Mbps, $\phi_{u_s} = 100$units/Mbps.

of all the schemes continue to grow with increasing values of α, ϕ, and γ, except the total utility of the distributed scheme without a caching function with respect to the value of γ. Furthermore, we can also observe from these figures that the gap between the proposed scheme and the centralized scheme is narrow.

4.6 Conclusion and future work

In this chapter, we proposed a novel framework with information-centric wireless HetNets. In the framework, we studied virtual resource allocation for communication, computing, and caching. The allocation strategy was formulated as a joint optimization problem, considering the gains of not only virtualization but also caching and computing in the information-centric HetNets virtualization architecture. In addition, a distributed ADMM-based algorithm was introduced to decouple the coupling variables and then split the optimization problem into several subproblems, since signaling overhead and computing complexity are prohibitively high in the centralized algorithm. Simulations were presented to show the convergence of the proposed distributed algorithm and the performance improvement of the proposed scheme. Future work is in progress to consider software-defined networking (SDN) in the proposed framework.

References

[1] Y. Zhou, F. R. Yu, J. Chen, and Y. Kuo, "Resource allocation for information-centric virtualized heterogeneous networks with in-network caching and mobile edge computing," *IEEE Transa. on Veh. Tech.*, vol. 66, no. 12, pp. 11339-11351, Dec. 2017.

[2] L. Wei, J. Cai, C. H. Foh, and B. He, "QoS-aware resource allocation for video transcoding in clouds," *IEEE Transa. Circuits Sys. for Video Technol.*, vol. 27, no. 1, pp. 49-61, Jan. 2017.

[3] W. Zhang, Y. Wen, and H.-H. Chen, "Toward transcoding as a service: Energy-efficient offloading policy for green mobile cloud," *IEEE Network*, vol. 28, no. 6, pp. 67-73, Dec. 2014.

[4] N. Kumar, S. Zeadally, and J. J. Rodrigues, "Vehicular delay-tolerant networks for smart grid data management using mobile edge computing," *IEEE Commun. Mag.*, vol. 54, no. 10, pp. 60-66, Oct. 2016.

[5] K. Zhang, Y. Mao, S. Leng, Q. Zhao, L. Li, X. Peng, L. Pan, S. Maharjan, and Y. Zhang, "Energy-efficient offloading for mobile edge computing in 5g heterogeneous networks," *IEEE Access*, vol. 4, pp. 5896-5907, 2016.

[6] X. Chen, L. Jiao, W. Li, and X. Fu, "Efficient multi-user computation offloading for mobile-edge cloud computing," *IEEE/ACM Trans. Netw.*, vol. 24, no. 5, pp. 2795-2808, Oct. 2016.

[7] Y. Zhou, F. R. Yu, J. Chen, and Y. Kuo, "Video transcoding, caching, and multicast for heterogeneous networks over wireless network virtualization," *IEEE Commun. Lett.*, vol. 22, no. 1, pp. 141-144, Jan. 2018.

[8] G. Xylomenos, C. N. Ververidis, V. A. Siris, N. Fotiou, C. Tsilopoulos, X. Vasilakos, K. V. Katsaros, and G. C. Polyzos, "A survey of information-centric networking research," *IEEE Commun. Surveys Tutorials*, vol. 16, no. 2, pp. 1024-1049, 2014.

[9] C. Fang, F. R. Yu, T. Huang, J. Liu, and Y. Liu, "A survey of green information-centric networking: Research issues and challenges," *IEEE Commun. Surveys Tutorials*, vol. 17, no. 3, pp. 1455-1472, 2015.

[10] C. Liang, F. R. Yu, and X. Zhang, "Information-centric network function virtualization over 5G mobile wireless networks," *IEEE Network*, vol. 29, no. 3, pp. 68-74, Jun. 2015.

[11] C. Liang and F. R. Yu, "Wireless network virtualization: A survey, some research issues and challenges," *IEEE Commun. Surveys Tutorials*, vol. 17, no. 1, pp. 358-380, 2015.

[12] K. Wang, F. R. Yu, and H. Li, "Information-centric virtualized cellular networks with device-to-device (D2D) communications," *IEEE Trans. Veh. Tech.*, vol. 65, no. 11, pp. 9319-9329, Nov. 2016.

[13] C. Liang, F. R. Yu, H. Yao, and Z. Han, "Virtual resource allocation in information-centric wireless networks with virtualization," *IEEE Trans. Veh. Tech.*, vol. 65, no. 12, pp. 9902-9914, Dec. 2016.

[14] S. Boyd, N. Parikh, E. Chu, B. Peleato, and J. Eckstein, "Distributed optimization and statistical learning via the alternating direction method of multipliers," *Foundations and Trends R in Machine Learning*, vol. 3, no. 1, pp. 1-122, Jan. 2011.

[15] A. Belbekkouche, M. M. Hasan, and A. Karmouch, "Resource discovery and allocation in network virtualization," *IEEE Commun. Surveys Tutorials*, vol. 14, no. 4, pp. 1114-1128, 2012.

[16] Y. Cai, F. R. Yu, C. Liang, B. Sun, and Q. Yan, "Software defined device-to-device (D2D) communications in virtual wireless networks with imperfect network state information (NSI)," *IEEE Trans. Veh. Tech.*, vol. 65, no. 9, pp. 7349-7360, Sept. 2016.

[17] Q. Yan and F. R. Yu, "Distributed denial of service attacks in software defined networking with cloud computing," *IEEE Commun. Mag.*, vol. 53, no. 4, pp. 52-59, Apr. 2015.

[18] Q. Yan, F. R. Yu, Q. Gong, and J. Li, "Software-defined networking (SDN) and distributed denial of service (DDoS) attacks in cloud computing environments: A survey, some research issues, and challenges," *IEEE Commun. Surveys Tutorials*, vol. 18, no. 1, pp. 602-622, 2016.

[19] L. Cui, F. R. Yu, and Q. Yan, "When big data meets software-defined networking: SDN for big data and big data for SDN," *IEEE Network*, vol. 30, no. 1, pp. 58-65, Feb. 2016.

[20] Mobile-Edge Computing (MEC) Industry Initiative, "Mobile-edge computing. Introductory Technical White Paper," 2014.

[21] M. Maier, M. Chowdhury, B. P. Rimal, and D. P. Van, "The tactile internet: Vision, recent progress, and open challenges," *IEEE Commun. Mag.*, vol. 54, no. 5, pp. 138-145, May 2016.

[22] J. O. Fajardo, I. Taboada, and F. Liberal, "Improving content delivery efficiency through multi-layer mobile edge adaptation," *IEEE Network*, vol. 29, no. 6, pp. 40-46, Dec. 2015.

[23] Y. L. Lee, J. Loo, T. C. Chuah, and A. A. El-Saleh, "Fair resource allocation with interference mitigation and resource reuse for LTE/LTEA femtocell networks," *IEEE Trans. Veh. Tech.*, vol. 65, no. 10, pp. 8203-8217, Oct. 2016.

[24] H. Zhang, C. Jiang, N. C. Beaulieu, X. Chu, X. Wen, and M. Tao, "Resource allocation in spectrum-sharing OFDMA femtocells with heterogeneous services," *IEEE Trans. Commun.*, vol. 62, no. 7, pp. 2366-2377, Jul. 2014.

[25] X. Kang, R. Zhang, and M. Motani, "Price-based resource allocation for spectrum-sharing femtocell networks: A stackelberg game approach," *IEEE J. Sel. Areas Commun.*, vol. 30, no. 3, pp. 538-549, Apr. 2012.

[26] J. Li, H. Chen, Y. Chen, Z. Lin, B. Vucetic, and L. Hanzo, "Pricing and resource allocation via game theory for a small-cell video caching system," *IEEE J. Sel. Areas Commun.*, vol. 34, no. 8, pp. 2115-2129, Aug. 2016.

[27] M. Leinonen, M. Codreanu, and M. Juntti, "Distributed joint resource and routing optimization in wireless sensor networks via alternating direction method of multipliers," *IEEE Trans. Wireless Commun.*, vol. 12, no. 11, pp. 5454-5467, Nov. 2013.

[28] S. Boyd and L. Vandenberghe, *Convex Optimization*. Cambridge University Press, 2004.

[29] G. Liu, H. Ji, F. R. Yu, Y. Li, and R. Xie, "Energy-efficient resource allocation in full-duplex relaying networks," in *2014 IEEE International Conference on Communications (ICC)*. IEEE, Jun. 2014, pp. 2400-2405.

[30] X. Li, N. Zhao, Y. Sun, and F. R. Yu, "Interference alignment based on antenna selection with imperfect channel state information in cognitive radio networks," *IEEE Trans. Veh. Tech.*, vol. 65, no. 7, pp. 5497-5511, Jul. 2016.

Chapter 5

Network Slicing and Caching in 5G Cellular Networks

Network slicing has been considered one of the key technologies in next generation mobile networks (5G), which can create a virtual network and provide customized services on demand. Most of the current work on network slicing focuses on virtualization technology, especially in virtual resource allocation. However, caching as a significant approach to improve content delivery and Quality of Experience (QoE) for end-users has not been well considered in network slicing. In this chapter, we consider in-network caching combines with network slicing, and propose an efficient caching resource allocation scheme for network slicing in a 5G core network. We first formulate the caching resource allocation issue as an integer linear programming (ILP) model, and then propose a caching resource allocation scheme based on a Chemical Reaction Optimization (CRO) algorithm, which can significantly improve caching resource utilization. The CRO algorithm is a population-based optimization metaheuristic, which has advantages in searching optimal solution and computation complexity. Finally, extensive simulation results are presented to illustrate the performance of the proposed scheme.

5.1 Introduction

To cope with both the challenges and opportunities brought by the rapid development of mobile network services and applications, such as high-definition video, virtual reality, online gaming, cloud services and so on, the fifth generation (5G) mobile network has been proposed and studied widely from the perspective of the innovation of network architecture and key technologies, respectively [1][2]. For 5G network architecture, one of the most important characteristics is network slicing introduced [3] [4] by using virtualization technology [5, 6, 7, 8, 9]. In this case, based on a single physical network infrastructure, the virtual network can be abstracted and created to provide end-to-end network services on demand for various use cases (e.g. smartphones, autonomous driving, massive Internet of Things (IoT) and so on), which can effectively save capital expenses (CapEx) and operation expenses (OpEx) for telecom operators.

For network slicing in 5G, one of the most important issues is how to flexibly and efficiently configure and manage network resources for a virtual network, which has attracted a lot of attention in recent years [10, 11, 12, 13, 14]. The problems of virtual network resource allocation and management in 5G networks are studied in [10], and a virtual link mapping scheme based on mixed integer programming is proposed. The authors in [11] propose a wireless network virtualization model, and in this model the infrastructure and resources are dynamically and flexibly sliced into multiple virtual networks, achieving optimal resource utilization. The authors of [12] study the future mobile core network, and then a 5G core network framework based on network slicing is proposed, which can enable cost-effective service deployment and efficient service operation. In [13], the authors propose a future oriented wireless network architecture, which considers network slicing and virtual resource allocation issues as well as in-network caching policy. The authors in [14] study the problem of two-level hierarchical resource allocation for 5G network slicing, and a hierarchical combinatorial auction mechanism is proposed to solve the hierarchical resource allocation problem.

On the other hand, with the extensive research of another new technology, information-centric networking (ICN), in recent years [15][16], in-network caching has been considered a promising technology in 5G network [17, 18, 19, 20] to reduce duplicate content transmission and improve the QoE of end-users. The authors of [17] study the potential caching techniques in 5G networks, and analyze the impact of caching deployment position on network performance. Moreover, the authors propose and verify a mobile network caching scheme based on content-centric networking [21]. That in-network caching can potentially help reduce user content access delay and mobile traffic load is demonstrated. The authors in [18] investigate the mobile network caching problem, and jointly consider the content placement issue and the request routing issue. Thus, a collaborative caching algorithm is proposed to save network bandwidth cost as well as improve the QoE of subscribers. In

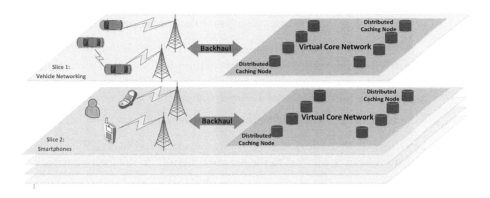

Figure 5.1: 5G network slices with virtual distributed node caching [3].

order to enhance video content delivery in mobile core networks, the authors of [19] propose a collaborative caching framework, and formulate the in-network caching problem as a network total cost minimizing issue. In [20], the authors present a caching virtualization scheme for a mobile network, and emphasize the significance of caching as a kind of network service.

In [22], the authors study the peak rate of the caching problem under the scenario that users are equipped with small buffer sizes, and then propose a novel coded caching strategy to reduce the peak rate. In [23], the authors study the problem of downlink energy saving in a video sharing scenario and propose two approaches, namely, D2D cluster and D2D caching, under the coalition game framework.

Although some excellent work has been done on network slicing and caching in 5G networks, to the best of our knowledge, these two significant issues have traditionally been solved separately in the literature. However, as described in the following, it is necessary to jointly consider these two advanced technologies to satisfy performance requirements and improve resource utilization in 5G networks. Therefore, in this chapter, involving an in-network caching feature in network slicing, we propose to jointly consider in-network caching and network slicing to improve end-to-end system performance. Note that we mainly focus on these two technologies in a 5G core network. The motivations of our work are based on the following considerations.

■ Integrating in-network caching and network slicing in a 5G core network is an important network evolution trend and has important significance, which cannot only reduce the mobile network traffic load and improve the performance of content delivery, but also flexibly provide various customized services on demand.

■ In the existing works of in-network caching and network slicing, in-network caching resource allocation has not been well considered in network slicing, which may result in suboptimal performance of network slicing.

■ Accordingly, the scheme of efficient caching resource allocation for network slicing in a 5G core network has the potential to significantly improve the performance of network slicing and QoE of end-users.

The main contributions of this chapter are summarized as follows:

■ By integrating network slicing and in-network caching, the problem of caching resource allocation for network slicing in a 5G core network is studied.

■ We mainly focus on maximizing the utilization of physical caching resources for network slicing via price mechanism, then we formulate the caching resource allocation issue as an integer linear programming model. Particularly, we also take caching energy consumption into account due to the importance of caching energy efficiency in a 5G network [24, 25, 26].

■ The Chemical Reaction Optimization (CRO) [27] algorithm based caching resource allocation scheme for network slicing is proposed to maximize caching resource utilization. The CRO algorithm is a population-based optimization metaheuristic, mimicking the molecular interactions in a chemical reaction, with advantages in searching optimal solution and computation complexity.

Finally, extensive simulation results are presented to illustrate the effectiveness of the proposed scheme. The rest of this chapter is organized as follows. In Section 5.2, an overview of 5G core network integrating network slicing and caching is presented first, and then the system model and problem formulation for the caching resource allocation problem are given. In Section 5.3, a CRO-based caching resource allocation scheme is proposed to solve the caching resource allocation problem. In Section 5.4, we evaluate the performance of the CRO-based scheme. We conclude and describe future work in Section 5.5.

5.2   System model and problem formulation

In this section, we first briefly present an overview of a 5G core network involving network slicing and caching. Then, the system model and problem formulation of caching resource allocation for network slicing in a 5G core network are described.

5.2.1 Overview of a 5G core network involving network slicing and caching

For network slicing in a 5G network, physical network infrastructure resources are abstracted and sliced into virtual network resources, which can hold certain corresponding functionalities. In addition, these virtual networks that are usually called network slices can provide various network services [3] [4], as shown in Fig 5.1. For example, the network slice for typical smartphones use can provide mobile network data and voice services as the current mobile cellular network. And the network slice for vehicle networking use can provide multimedia service and autonomous driving service. Moreover, many network slices require low latency to satisfy the demand of service, such as autonomous driving service, mobile high-definition video service, etc. In order to improve the efficiency of content delivery and QoE for end-users, deploying distributed in-network caching in a 5G core network is a significant approach [17]. As shown in Fig 5.2, each caching node adopts virtualization technology, which is abstracted as a virtual caching resource pool and dynamically provides content cache for 5G network slices. Particularly, in this model we do not consider the mapping mechanism and admission control between the network slices and the physical network. We assume the mapping results are already known and network slices are allowed to access the physical network resources[4][28][29].

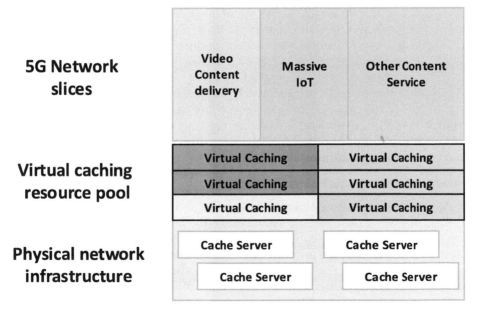

Figure 5.2: The caching node with virtualization technology.

5.2.2   System model and problem formulation

In this subsection, we introduce the system model and problem formulation. In this chapter, we assume that the infrastructure providers (InPs) own and operate physical network infrastructure resources. And InPs can slice physical network resources into multiple virtual networks or network slices, and lease these network slices to the service providers (SPs) according to a certain price mechanism. Then SPs rent the virtual network resources from InPs relying on its demand, and provide various services for their customers.

In this system model, we assume that the network slice is a collection of virtual nodes and virtual links toward customized services, which is created by InP and operated by SP. Note that each physical network node is integrated with a caching resource in this model. In this chapter, we only consider one InP and multiple SPs, and each SP only rents a network slice. Hence, it is an important issue how to allocate the virtual resources among multiple network slices. Particularly, we only consider caching resource allocation in this model. It is obvious that the allocation of caching resources has the most direct influence on the performance of each network slice and the entire 5G network. Moreover, we transform the caching resource allocation optimization issue into maximizing the revenue problem for InP. Based on the above assumptions, we can build the problem formulation of caching resource allocation as follows.

Here, we assume that the physical network is managed by the only InP, and the physical network can be denoted as a weighted undirected graph as $\mathcal{G} = (\mathcal{N}, \mathcal{L})$, where $\mathcal{N}$ is the set of physical nodes denoted by $\mathcal{N} = \{1, 2, 3, ..., N\}$, and $\mathcal{L}$ is the set of physical links. In addition, we assume the caching resource capacity of each physical node is limited, and $\mathcal{C} = \{c_1, c_2, ..., c_N\}$ denotes the caching resource capacity on each physical node, and the caching resource capacity of node i is expressed as c_i.

The network slices can be denoted by $\mathcal{V} = \{1, 2, ..., M\}$, which means that M network slices coexist in this model. We first introduce an integer variable $x_{ik} \in \{0, 1\}$. And $x_{ik} = 1$ means the caching resource on physical node i is allocated to the network slice k within the unit time T, otherwise $x_{ik} = 0$. Meanwhile, the caching resource capacity of the physical node i occupied by the network slice k in unit time T denotes $y_{x_{ik}}$, which naturally follows the constraint of $\forall 1 \leq i \leq N, 1 \leq k \leq M, \sum_{k=1}^{M} x_{ik} y_{x_{ik}} \leq c_i$.

Moreover, the SP as a tenant leases the caching resource from the InP. Hence, the SP should pay a certain charge to the InP. The prices of caching resources are usually determined by both SP and InP at the time of collaboration. We assume that the prices of caching resources in different network nodes are different, because the importance of each network node is different[30]. For example, a node is located in the hub position of a network, and we think this node is more important. Moreover, the importance of nodes may need to be negotiated and an agreement reached between the InP and SP. Hence, we can set the weight for each node, and the weight of node i can be denoted as w_i.

We assume the price of caching resource per unit capacity is denoted by ρ, thus the price-weighted caching resource per unit capacity in node i can be expressed as $p_i = \rho w_i$.

Based on the above discussion, we can find that the network slice k should pay $p_i x_{ik} y_{x_{ik}}$ for node i in unit time T. Correspondingly, the network slice k should pay P_k for all the nodes of InP, which can be written as:

$$P_k = \sum_{i=1}^{N} p_i x_{ik} y_{x_{ik}} T \tag{5.1}$$

Therefore, all the network slices should pay P for InP, which can be written as:

$$P = \sum_{k=1}^{M} P_k = \sum_{k=1}^{M} \sum_{i=1}^{N} \rho w_i x_{ik} y_{x_{ik}} T \tag{5.2}$$

Providing caching services will bring a certain cost to InP. In many studies, energy consumption is considered a cost, and many energy consumption models have been proposed. In reference [31][32], the authors develop a power consumption model for an ICN router from the perspective of hardware platform, which mainly includes the power consumption of the CPU device, memory device and network interface Card. In this chapter, we mainly consider the caching energy consumption, which mainly consists of content caching and content responding[33][34]. Hence, we can give the energy consumption cost model regarding content caching and content responding as follows:

Energy consumption cost regarding content caching, P_{cache}: The energy consumption cost regarding content caching mainly includes the energy consumption of content caching and content updating. Hence, we can build an energy-proportional model. And the energy consumption can be expressed as:

$$E_{cache} = \sum_{k=1}^{M} \sum_{i=1}^{N} \alpha x_{ik} y_{x_{ik}} v_{ik} T \tag{5.3}$$

where α is the value of energy consumption for content caching or content updating per unit cache capacity, and v_{ik} is the content update time (or update frequency) in unit time T. Particularly, in order to avoid caching node overload, we can set an update limit V, which is the maximum of content update times, namely, $v_{ik} \leq V$. Hence, the energy consumption cost can be written as:

$$P_{cache} = \rho_c E_{cache} = \rho_c \sum_{k=1}^{M} \sum_{i=1}^{N} \alpha x_{ik} y_{x_{ik}} v_{ik} T \tag{5.4}$$

where ρ_c is the price of energy consumption for content caching or content updating per unit cache capacity.

Energy consumption cost regarding content response, P_{res}: The energy consumption cost regarding content response mainly comes from the energy consumption of responding and serving user content requests. The energy consumption can be expressed as:

$$E_{res} = \sum_{k=1}^{M} \sum_{i=1}^{N} \beta x_{ik} y_{x_{ik}} \gamma_{ik} T \tag{5.5}$$

where β is the value of energy consumption for content response per unit cache capacity, and γ_{ik} is the content response time (or content response frequency) in unit time T. Particularly, in order to avoid caching node overload, we can set a response limit R, which is the maximum of content response times, namely, $\gamma_{ik} \leq R$. Hence, the energy consumption cost can be written as:

$$P_{res} = \rho_r E_{res} = \rho_r \sum_{k=1}^{M} \sum_{i=1}^{N} \beta x_{ik} y_{x_{ik}} \gamma_{ik} T \tag{5.6}$$

where ρ_r is the price of energy consumption for content response per unit cache capacity. Hence, the total cost of energy consumption can be written as:

$$P_{cost} = P_{cache} + P_{res} = \rho_c E_{cache} + \rho_r E_{res} \tag{5.7}$$

Based on the above description, this optimization problem can be converted into maximizing the InP total revenue by optimizing caching resource allocation, and the total revenue function of InP can be defined as U_{total}, which can be written as follows:

$$\max \quad U_{total} = P - P_{cost} = P - P_{cache} - P_{res} \tag{5.8}$$

Here, we can use a two-dimensional matrix $\mathbf{Y} \in \mathcal{Y}$ with size $N \times M$ to denote the state of caching resource allocation for network slices. Obviously, the two-dimensional matrix $\mathbf{Y}$ is a solution to the problem of caching resource allocation. And $\mathcal{Y}$ denotes the set of all possible caching resource allocations. Hence, the total revenue U_{total} can be rewritten as:

$$U_{total}(\mathbf{Y}) = \sum_{k=1}^{M} \sum_{i=1}^{N} \rho w_i x_{ik} y_{x_{ik}} T - \rho_c \sum_{k=1}^{M} \sum_{i=1}^{N} \alpha x_{ik} y_{x_{ik}} v_{ik} T - \rho_r$$
$$\times \sum_{k=1}^{M} \sum_{i=1}^{N} \beta x_{ik} y_{x_{ik}} \gamma_{ik} T \tag{5.9}$$

Now, the aggregate utility maximization problem is shown as follows:

$$\max \quad U_{total}(\mathbf{Y}) \tag{5.10}$$

$$Subject \quad to:$$
$$\sum_{k=1}^{M} x_{ik} y_{x_{ik}} \leq c_i, \quad \forall i \in N, \quad \forall k \in M$$
$$w_i > 0, \quad \forall i \in N$$
$$\rho_c > 0, \rho_r > 0, \rho > \rho_c, \rho > \rho_r \tag{5.11}$$
$$0 < \gamma_{ik} \leq R, \quad 0 < v_{ik} \leq V$$
$$x_{lk} \subset \{0, 1\}$$

Table 5.1: Key Notations

variable	description
N	Number of physical caching nodes
M	Number of network slices
x_{ik}	An indicator that node i allocates caching resource to slice k
$y_{x_{ik}}$	Caching resource allocation matrix
c_i	Caching capacity of the physical caching node i
ρ	the price of caching resource per unit capacity
w_i	the weight of the physical caching node i
ρ_c	The price of energy consumption for content caching or content updating per unit cache capacity
ρ_r	The price of energy consumption for content response per unit cache capacity
α	The value of energy consumption for content caching or content updating per unit cache capacity
β	The value of energy consumption for content response per unit cache capacity
υ_{ik}	The content update frequency in unit time T
γ_{ik}	The content response frequency in unit time T

The first constraint denotes that the caching resource allocated to network slices is limited for each caching node. The second constraint denotes that price-weighting is valid. The key notations are listed in table 5.1.

5.3 Caching resource allocation based on the CRO algorithm

In this section, we propose a caching resource allocation scheme based on the CRO algorithm. Efficient in-network caching can be mapped to the knapsack problem, which is an NP-hard problem [35]. Hence, efficient in-network caching can be solved by a heuristic algorithm. Here, from the optimization problem (5.10) in Subsection 5.2.2, we find that solving the caching resource allocation matrix $\mathbf{Y}$ is NP-hard. Hence, we use CRO, a metaheuristic algorithm, to solve the optimization problem, thus obtaining a suboptimal solution. Here, we first describe the CRO algorithm, and then the caching resource allocation scheme based on the CRO algorithm is proposed.

5.3.1   Brief introduction to the CRO algorithm

The CRO algorithm is a population-based metaheuristic for optimization, which is inspired by the characteristic of a chemical reaction process. The CRO algorithm mimics the molecule actions in a chemical reaction process, which can reach a steady state with low energy [27]. Although the CRO is a recently proposed optimization algorithm, it has achieved great success. Many cases or problems can utilize the CRO, such as energy-efficient content caching [36], cognitive radio resource allocation [37], task scheduling in grid computing [38], etc.

In the CRO algorithm, the main research objects are molecules, and each molecule mainly consists of molecular structure and molecular energy. The molecular energy includes potential energy (PE) as well as kinetic energy (KE). Corresponding to a given optimization problem, the molecular structure represents the solution of the given problem, and the PE represents the objective function value of the problem. The KE means the tolerance of a molecule getting a higher potential energy state than the existing one, thus allowing the CRO to escape from local optimal solutions [27]. Hence, the core idea of the CRO algorithm is the molecules try to reach the stable state with the minimum potential energy by a chemical reaction process. Namely, after the end of the chemical reaction, the molecules can naturally reach a steady state, and the potential energy of molecules correspondingly becomes the lowest.

During the process of a chemical reaction, a series of molecular collisions occurs, and the molecules collide either with each other or with the walls of the container. Hence, in the CRO algorithm, four types of molecular actions are considered, which include on-wall ineffective collision, decomposition, intermolecular ineffective collision, and synthesis. Moreover, different degrees of molecular actions can lead to varying degrees of energy change. Corresponding to the optimization problem, they have different characteristics and extent of change to the solutions. Therefore, through a series of chemical reactions, we can obtain the molecular structure with the lowest energy states of PE, then we can output it as the best solution of the given problem [37].

5.3.2   Caching resource allocation based on the CRO algorithm

In this subsection, we propose a caching resource allocation scheme based on the CRO algorithm. As pointed out in Section II, the caching resource allocation matrix $\mathbf{Y}$ is the solution of the given optimization issue. Hence, we can optimize the caching assignment matrix $\mathbf{Y}$ to obtain the optimal caching resource allocation. At first, the InP randomly selects network slices to allocate caching resources under the constraint condition, and the matrix $\mathbf{X}$ is a 0/1 indicator matrix, specifying which network slices are assigned to caching resources. Matrix $\mathbf{Y}$ is a caching resource allocation matrix, specifying how many of the cache resources are assigned to the network slices. Therefore, we want to find an optimal caching resource allocation matrix **Y** to maximize the objective function. In order to obtain the optimal caching resource allocation matrix **Y**, we use the CRO algorithm.

Here, the whole network can be regard as a molecule. Correspondingly, the molecular structure represents the solution of the given optimization issue, namely, the caching resource allocation matrix **Y**. The PE represents the objective function value, namely, the total InP revenue.

Hence, the flow chart of the caching allocation scheme based on the CRO algorithm is shown in Fig 5.3. A more detailed illustration for the caching resource allocation scheme based on the CRO algorithm can be described as follows:

1) *Initialization:* we can generate a group of initial solutions in the first step. In detail, we first set the initial parameters of the CRO, such as $PopSize, KELossRate, InitialKE, MoleColl$ [27]. Then, we randomly generate indicator matrix **X** under the constraint condition, specifying which network slices are allocated to the nodes' caching resources. Next, on the basis of indicator matrix **X**, we randomly generate caching resources allocation matrix **Y** under the constraint condition, specifying how many of the cache resources are occupied by the network slices.

2) *On-wall Ineffective Collision:* In a chemical reaction, the collision between a molecule and a wall of container has a slight effect on the molecule structure. Correspondingly, in this step of the proposed algorithm, we can obtain a new solution with a slight change. Hence, we can choose one caching node and change the allocation of caching resources. Specifically, we can randomly select one row of data from the caching allocation matrix **Y**, then we randomly reallocate caching resources under constraint conditions. Hence, we get a new caching resource allocation matrix **Y**′. For example,

$$
\underbrace{\begin{bmatrix} a_{11} & a_{12} & a_{13} & a_{14} \\ a_{21} & a_{22} & a_{23} & a_{24} \\ a_{31} & a_{32} & a_{33} & a_{34} \\ a_{41} & a_{42} & a_{43} & a_{44} \end{bmatrix}}_{Y} \rightarrow \underbrace{\begin{bmatrix} a_{11} & a_{12} & a_{13} & a_{14} \\ \xi_{21} & \xi_{22} & \xi_{23} & \xi_{24} \\ a_{31} & a_{32} & a_{33} & a_{34} \\ a_{41} & a_{42} & a_{43} & a_{44} \end{bmatrix}}_{Y'}
$$

3) *Decomposition:* In this process of a chemical reaction, a molecule usually collides with the wall of container and splits into two new molecules. Compared with the original molecule, the new resultant molecules have huge changes. Hence, we can obtain two new solutions $\mathbf{Y}'_1$ and $\mathbf{Y}'_2$, each of which has a big difference from the original **Y**. In the proposed algorithm, we randomly select a row of caching allocation matrix **Y** and randomly assign it to the same row of either $\mathbf{Y}'_1$ or $\mathbf{Y}'_2$, until each row of the original matrix is assigned to the two new matrices. For those rows no assigned values in caching matrix $\mathbf{Y}'_1$ or $\mathbf{Y}'_2$, we randomly assign a positive value to them under the constraint condition.

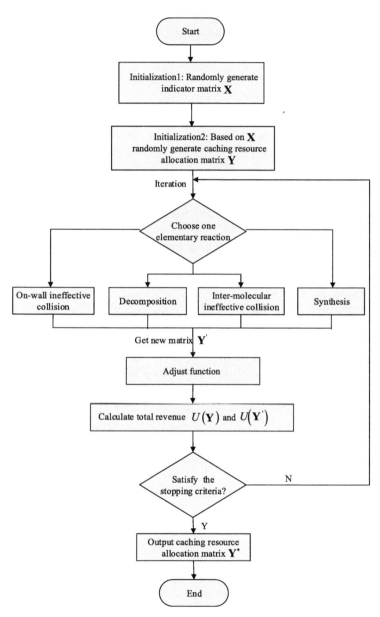

Figure 5.3: Flow chart for the CRO-based caching resource allocation.

Therefore, we obtain two new caching resource allocation matrices $\mathbf{Y}'_1$ and $\mathbf{Y}'_2$. For example,

$$
\underbrace{\begin{bmatrix} a_{11} & a_{12} & a_{13} & a_{14} \\ a_{21} & a_{22} & a_{23} & a_{24} \\ a_{31} & a_{32} & a_{33} & a_{34} \\ a_{41} & a_{42} & a_{43} & a_{44} \end{bmatrix}}_{Y} \longrightarrow
$$

$$
\underbrace{\begin{bmatrix} a_{11} & a_{12} & a_{13} & a_{14} \\ \underline{\xi_{21}} & \underline{\xi_{22}} & \underline{\xi_{23}} & \underline{\xi_{24}} \\ a_{31} & a_{32} & a_{33} & a_{34} \\ \underline{\xi_{41}} & \underline{\xi_{42}} & \underline{\xi_{43}} & \underline{\xi_{44}} \end{bmatrix}}_{Y'_1} \; and \; \underbrace{\begin{bmatrix} \xi_{11} & \xi_{12} & \xi_{13} & \xi_{14} \\ a_{21} & a_{22} & a_{23} & a_{24} \\ \xi_{31} & \xi_{32} & \xi_{33} & \xi_{34} \\ a_{41} & a_{42} & a_{43} & a_{44} \end{bmatrix}}_{Y'_2}
$$

4) *Inter-molecular Ineffective Collision:* In this step of a chemical reaction, two molecules collide with each other and the collision between the two molecules has a slight effect on each molecules structure. Thus, the change of each molecules structure in this step is similar to that in on-wall ineffective collision. Correspondingly, we can get two new caching allocation matrices $\mathbf{Y}'_1$ and $\mathbf{Y}'_2$ from the original caching allocation matrices $\mathbf{Y}_1$ and $\mathbf{Y}_2$, and each caching allocation matrix has a slight change compared to its original matrix. Hence, both $\mathbf{Y}_1$ and $\mathbf{Y}_2$ can adopt the method used in the on-wall ineffective collision to reallocate caching resources separately. For example,

$$
\underbrace{\begin{bmatrix} a_{11} & a_{12} & a_{13} & a_{14} \\ a_{21} & a_{22} & a_{23} & a_{24} \\ \underline{a_{31}} & \underline{a_{32}} & \underline{a_{33}} & \underline{a_{34}} \\ a_{41} & a_{42} & a_{43} & a_{44} \end{bmatrix}}_{Y_1} \; and \; \underbrace{\begin{bmatrix} b_{11} & b_{12} & b_{13} & b_{14} \\ b_{21} & b_{22} & b_{23} & b_{24} \\ b_{31} & b_{32} & b_{33} & b_{34} \\ b_{41} & b_{42} & b_{43} & b_{44} \end{bmatrix}}_{Y_2} \longrightarrow
$$

$$
\underbrace{\begin{bmatrix} a_{11} & a_{12} & a_{13} & a_{14} \\ a_{21} & a_{22} & a_{23} & a_{24} \\ \underline{\xi_{31}} & \underline{\xi_{32}} & \underline{\xi_{33}} & \underline{\xi_{34}} \\ a_{41} & a_{42} & a_{43} & a_{44} \end{bmatrix}}_{Y'_1} \; and \; \underbrace{\begin{bmatrix} \xi_{11} & \xi_{12} & \xi_{13} & \xi_{14} \\ b_{21} & b_{22} & b_{23} & b_{24} \\ b_{31} & b_{32} & b_{33} & b_{34} \\ b_{41} & b_{42} & b_{43} & b_{44} \end{bmatrix}}_{Y'_2}
$$

5) *Synthesis:* In this process of a chemical reaction, two molecules collide with each other and combine to form a new molecule. Obviously, the new resultant molecule is quite different from the original two molecules. Correspondingly, the new resultant caching allocation matrix $\mathbf{Y}'$ should be quite different from the original two matrices $\mathbf{Y}_1$ and $\mathbf{Y}_2$. Here, we randomly select the values of corresponding rows from two matrices, and generate a new caching allocation matrix $\mathbf{Y}'$. For example,

$$\underbrace{\begin{bmatrix} a_{11} & a_{12} & a_{13} & a_{14} \\ a_{21} & a_{22} & a_{23} & a_{24} \\ a_{31} & a_{32} & a_{33} & a_{34} \\ a_{41} & a_{42} & a_{43} & a_{44} \end{bmatrix}}_{Y_1} and \underbrace{\begin{bmatrix} b_{11} & b_{12} & b_{13} & b_{14} \\ b_{21} & b_{22} & b_{23} & b_{24} \\ b_{31} & b_{32} & b_{33} & b_{34} \\ b_{41} & b_{42} & b_{43} & b_{44} \end{bmatrix}}_{Y_2}$$

$$\rightarrow \underbrace{\begin{bmatrix} b_{21} & b_{22} & b_{23} & b_{24} \\ a_{21} & a_{22} & a_{23} & a_{24} \\ a_{31} & a_{32} & a_{33} & a_{34} \\ b_{41} & b_{42} & b_{43} & b_{44} \end{bmatrix}}_{Y'}$$

6) *Adjust Function:* In this step of the proposed algorithm, we need to determine whether the solutions are satisfied to the constraints in (5.10). In particular, we focus on whether the new solution satisfies the constraint condition $\sum_{k=1}^{M} x_{ik} y_{x_{ik}} \leq c_i$ at the adjust function. If the new solution does not satisfy the constraint, we remove the caching resources from network slices randomly.

7) *Stopping Criteria:* At the final stage of the proposed algorithm, we have to check whether the stopping criterion is satisfied. Here, we set a "*for*" cycle and set the number of the cycle to θ. When the times of the "*for*" cycle reach θ, stop the iteration, otherwise, continue to repeat the iteration. After the iteration stops, we output the PE and molecular structure, namely, the objective function value and the solution of the given problem.

5.3.3   Complexity analysis

In this subsection, we analyze the algorithm complexity of the CRO-based algorithm. In the initialization stage, we need to generate the set of initial solutions, and this procedure is related to the number of initial solutions as well as the initial caching resource allocation matrix. Hence, from the procedure of the initialization stage, we find the time complexity is $O(PopSizeMN)$, where M is the number of network slices, N is the number of network nodes, and the $PopSize$ is the number of initial molecules, namely, the number of initial solutions. In the reaction process stage, we check the time complexity of the four elementary operations separately, and find the time complexity of one iteration in the process of the CRO-based algorithm. From the procedure of the CRO-based algorithm, we find that the time complexity of each elementary operation is $O(MN)$. Hence, the time complexity of the CRO-based algorithm is $O(IMN)$, where I is the number of iterations in the CRO-based algorithm. In the CRO-based algorithm, the optimization is mainly implemented by four elementary reaction operations. Each reaction operation can use the encapsulated function module, e.g., objective function module. Through a simple modification, the CRO-based algorithm can be easily applied to other optimization problems. Thus the CRO-based algorithm has good scalability. Moreover, it also can be exploited in a dynamic setting.

In the greedy-based algorithm, the time complexity of one iteration is also $O(MN)$. And the time complexity of the greedy-based algorithm is $O(IMN)$, where I is the number of iterations in the greedy-based algorithm. Hence, from the perspective of time complexity, the CRO-based algorithm and the greedy-based algorithm here have the same time complexity. In addition, the greedy algorithm also has good scalability and can be exploited in a dynamic setting. The greedy-based algorithm usually makes a locally optimal choice in each iteration. However, the CRO-based algorithm can escape from the locally optimal solution, and search the global optimal solution.

5.4   Simulation results and discussions

In this section, we evaluate the performance of the caching resource allocation scheme using the numerical simulation method. We consider a distributed 5G core network with caching nodes deploying the edge of the core network. To compare our proposed algorithm, three benchmarks are considered. The first is a random algorithm (RAND), which randomly allocates caching resources to network slices under constraint conditions. The second is the exhausting-based algorithm, which allocates all the caching resources to network slices under constraint conditions. The third is the greedy-based algorithm, which is a common method in caching allocation problems [39] [40]. Moreover, the greedy-based algorithm is a simple and efficient method, which always makes the choice that looks best at the moment. Namely, the greedy-based algorithm usually makes a locally optimal choice at each stage with the hope of finding a globally optimal solution.

In this simulation, we set the main parameter values of the CRO-based scheme as follows: $PopSize = 50$, $KELossRate = 0.2$, $InitialKE = 10000000$, $MoleColl = 0.5$. If there is no special explanation, the number of network nodes is generally set to 100 and the number of network slices is usually set to 50. We assume the node cache capacity is the same size and usually set to 100. Particularly, the caching resource allocation scheme in this chapter is a maximization issue. However, the CRO algorithm is designed for minimization issues. Hence, we need to modify the objective function as in [37] [41]. More specifically, we minimize $U'(Y) = 10^{10} - U(Y)$ rather than maximizing $U(Y)$ in the proposed scheme, and then we output $U(Y) = 10^{10} - U'(Y)$.

Fig 5.4 depicts the InP total revenue when the number of caching nodes increases from 100 to 1000. From Fig 5.4, we can see InP total revenue increases as the number of caching nodes increases. This is because InP can provide more caching resources for SPs with the increase of caching nodes, and thus SPs can be allocated more caching resources and pay more for InP. In additon, as shown in Fig 5.4, the CRO-based scheme can achieve a higher total

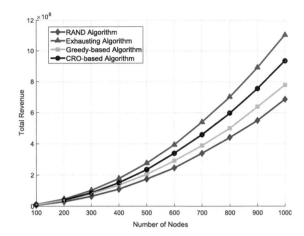

Figure 5.4: The total revenue versus the number of network nodes.

revenue than the RAND algorithm and greedy-based algorithm, and is more approximate to the exhausting-based algorithm, which means the proposed scheme is more approximate to the optimal solution.

Fig 5.5 shows the InP total revenue when the number of network slices increases from 10 to 60. In this case, the total revenue of the exhausting-based scheme is constant, as shown in Fig 5.5. This is because the number of network nodes and the capacity of each caching node are constant, thus total amount of caching resource is constant even if the network slices increase. Hence, the total revenue of the exhausting-based scheme is constant. However, when using the RAND algorithm, greedy-based algorithm and the CRO-based algorithm,

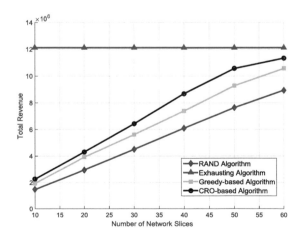

Figure 5.5: The total revenue versus the number of network slices.

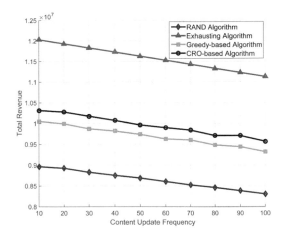

Figure 5.6: The total revenue versus the content update frequency.

the InP total revenue increases when the number of network slices increases, as shown in the figure. The reason is there are more caching resources allocated with the increase of network slices. Note that in this case, a parts of caching resources are still not allocated to network slices.

Fig 5.6 shows the performance of the total revenue when the content updating frequency varies from 10 to 100. As shown in Fig 5.6, the InP total revenue decreases with the increase of content updating frequency. The reason is the energy consumption of the caching node increases when the content updating frequency increases, thus increasing the energy consumption cost and decreasing the total revenue of InP.

Fig 5.7 depicts the performance of the total revenue when the content response frequency varies from 10 to 100. As shown in Fig 5.7, the InP total revenue decreases as the content response frequency increases. The reason is the energy consumption of the caching node increases as the content response frequency increases, thus increasing the energy consumption cost and decreasing the total revenue.

Fig 5.8 shows the InP total revenue when the size of the node caching capacity varies from 80 to 120. We assume that the size of each node caching capacity is the same in the model. In Fig 5.8, we observe the InP total revenue increases when the caching node capacity increases. The reason is there are more caching resources allocated to network slices with the increase of each nodes caching capacity, thus the SPs pay more for InP. Therefore, the total revenue of InP increases.

Fig 5.9 shows the energy consumption cost when the size of the node cache capacity varies from 80 to 120. In this simulation, we utilize the CRO-based algorithm to optimize caching resource allocation for minimizing energy consumption cost. We assume the size of each node caching capacity is the same in the model. As shown in Fig 5.9, the energy consumption cost increases

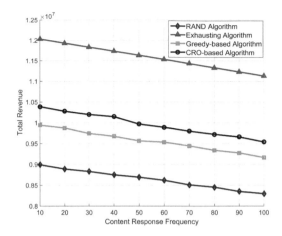

Figure 5.7: The total revenue versus the content response frequency.

when the caching node capacity increases. The reason is there are more content cached in the network node with the increase of each node caching capacity, thus resulting in more energy consumption of content caching and content response. Moreover, we observe that the CRO-based allocation scheme can save more energy consumption compared to the greedy-based algorithm and the RAND algorithm from Fig 5.9.

Fig 5.10 shows the energy consumption cost when the number of network slices varies from 10 to 60. In this simulation, the energy consumption cost of the exhausting-based scheme is constant. This is because the number of network nodes and the capacity of each caching node are constant, thus the

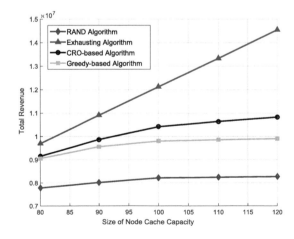

Figure 5.8: The total revenue versus the size of node cache capacity.

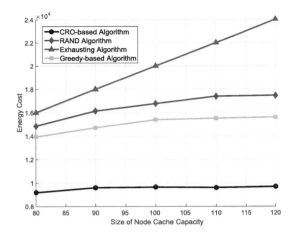

Figure 5.9: The energy consumption cost versus the size of node cache capacity.

total amount of caching resources is constant even if the network slices increase. Hence, the energy consumption cost of the exhausting-based scheme is constant. However, when using the RAND algorithm, greedy-based algorithm and the CRO-based algorithm, the energy consumption cost increases when

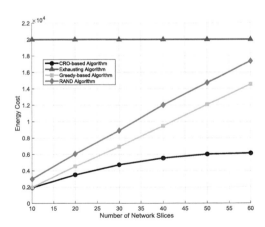

Figure 5.10: The energy consumption cost versus the number of network slices.

the number of network slices increases, as shown in the figure. The reason is there are more caching resources allocated with the increase of network slices. Note that in this case, a part of caching resources is still not allocated to network slices. From this figure, we find the CRO-based algorithm can save more energy consumption, which means the CRO-based algorithm has better performance.

5.5   Conclusions and future work

Integrating network slicing and in-network caching in a 5G core network has been considered a significant trend. One of the most important research issues is caching resource allocation to improve utilization of physical caching and save the cost of CapEx and OpEx. In this chapter, we have studied the issue of caching resource allocation for network slicing in the 5G core network. Moreover, we have formulated the caching resource allocation problem as an ILP model to maximize the total revenue of InP, and then proposed a caching resource allocation scheme based on the CRO algorithm. Finally, simulation results have shown that the CRO-based caching resource allocation scheme has better performance. In the future, we will study the issues of caching resource allocation in the 5G edge network.

References

[1] A. Gupta and R. K. Jha, "A survey of 5G network: Architecture and emerging technologies," *IEEE Access*, vol. 3, pp. 1206–1232, 2015.

[2] N. Panwar, S. Sharma, and A. K. Singh, "A survey on 5g: The next generation of mobile communication," *Physical Commun.*, 2015.

[3] N. Alliance, "5G white paper," *Next Generation Mobile Networks, White paper*, 2015.

[4] X. Zhou, R. Li, T. Chen, and H. Zhang, "Network slicing as a service: enabling enterprises' own software-defined cellular networks," *IEEE Commun. Mag.*, vol. 54, no. 7, pp. 146–153, July 2016.

[5] P. K. Agyapong, M. Iwamura, D. Staehle, W. Kiess, and A. Benjebbour, "Design considerations for a 5G network architecture," *IEEE Commun. Mag.*, vol. 52, no. 11, pp. 65–75, Nov 2014.

[6] P. Rost, I. Berberana, A. Maeder, H. Paul, V. Suryaprakash, M. Valenti, D. Wbben, A. Dekorsy, and G. Fettweis, "Benefits and challenges of virtualization in 5G radio access networks," *IEEE Commun. Mag.*, vol. 53, no. 12, pp. 75–82, Dec 2015.

[7] J. Costa-Requena, J. L. Santos, V. F. Guasch, K. Ahokas, G. Premsankar, S. Luukkainen, O. L. Prez, M. U. Itzazelaia, I. Ahmad, M. Liyanage, M. Ylianttila, and E. M. de Oca, "SDN and NFV integration in generalized mobile network architecture," in *Proc. IEEE EuCNC'15, Paris, France*, June 2015, pp. 154–158.

[8] H. Hawilo, A. Shami, M. Mirahmadi, and R. Asal, "NFV: state of the art, challenges, and implementation in next generation mobile networks (vEPC)," *IEEE Network*, vol. 28, no. 6, pp. 18–26, Nov 2014.

[9] F. Z. Yousaf, J. Lessmann, P. Loureiro, and S. Schmid, "SoftEPC-Dynamic instantiation of mobile core network entities for efficient resource utilization," in *Proc. IEEE ICC'13, Budapest, Hungary*, June 2013, pp. 3602–3606.

[10] R. Trivisonno, R. Guerzoni, I. Vaishnavi, and A. Frimpong, "Network Resource Management and QoS in SDN-Enabled 5G Systems," in *Proc.IEEE GLOBECOM'15, San Diego, CA, USA*, Dec 2015, pp. 1–7.

[11] Z. Feng, C. Qiu, Z. Feng, Z. Wei, W. Li, and P. Zhang, "An effective approach to 5g: Wireless network virtualization," *IEEE Commun. Mag.*, vol. 53, no. 12, pp. 53–59, Dec 2015.

[12] T. Shimojo, Y. Takano, A. Khan, S. Kaptchouang, M. Tamura, and S. Iwashina, "Future mobile core network for efficient service operation," in *Proc. IEEE NetSoft'15, London, UK*, April 2015, pp. 1–6.

[13] C. Liang and F. R. Yu, "Virtual resource allocation in information-centric wireless virtual networks," in *Proc. IEEE ICC'15, London, UK*, June 2015, pp. 3915–3920.

[14] K. Zhu and E. Hossain, "Virtualization of 5G cellular networks as a hierarchical combinatorial auction," *IEEE Trans. Mobile Computing*, vol. PP, no. 99, pp. 1–1, 2015.

[15] G. Xylomenos, C. N. Ververidis, V. A. Siris, N. Fotiou, C. Tsilopoulos, X. Vasilakos, K. V. Katsaros, and G. C. Polyzos, "A Survey of Information-Centric Networking Research," *IEEE Commun. Surv. Tutorials*, vol. 16, no. 2, pp. 1024–1049, Second 2014.

[16] L. Zhang, A. Afanasyev, J. Burke, V. Jacobson, K. Claffy, P. Crowley, C. Papadopoulos, L. Wang, and B. Zhang, "Named data networking," *ACM Sigcomm Comput. Commun. Rev.*, vol. 44, no. 3, pp. 66–73, 2014.

[17] X. Wang, M. Chen, T. Taleb, A. Ksentini, and V. Leung, "Cache in the air: exploiting content caching and delivery techniques for 5g systems," *IEEE Commun. Mag.*, vol. 52, no. 2, pp. 131–139, February 2014.

[18] S. Ren, T. Lin, W. An, Y. Li, Y. Zhang, and Z. Xu, "Collaborative EPC and RAN Caching Algorithms for LTE Mobile Networks," in *Proc. IEEE GLOBECOM'15, San Diego, CA, USA*, Dec 2015, pp. 1–6.

[19] J. He, H. Zhang, B. Zhao, and S. Rangarajan, "A collaborative framework for in-network video caching in mobile networks," *arXiv preprint arXiv:1404.1108*, 2014.

[20] X. Li, X. Wang, C. Zhu, W. Cai, and V. C. M. Leung, "Caching-as-a-service: Virtual caching framework in the cloud-based mobile networks," in *Proc. IEEE INFOCOM'15, Hong Kong, China*, April 2015, pp. 372–377.

[21] V. Jacobson, D. K. Smetters, J. D. Thornton, M. F. Plass, N. H. Briggs, and R. L. Braynard, "Networking named content," in *Proc. 5th Int'l Conf. Emerging Networking Experiments and Technologies(CoNEXT'09)*. ACM, 2009, pp. 1–12.

[22] Z. Chen, P. Fan, and K. B. Letaief, "Fundamental limits of caching: Improved bounds for small buffer users," *Eprint Arxiv*, vol. 10, 2014.

[23] Y. Shen, C. Jiang, T. Q. S. Quek, and Y. Ren, "Device-to-device-assisted communications in cellular networks: An energy efficient approach in downlink video sharing scenario," *IEEE Transactions on Wireless Communications*, vol. 15, no. 2, pp. 1575–1587, Feb 2016.

[24] M. Savi, O. Ayoub, F. Musumeci, Z. Li, G. Verticale, and M. Tornatore, "Energy-efficient caching for video-on-demand in fixed-mobile convergent networks," in *IEEE OnlineGreenComm'15*, Nov 2015, pp. 17–22.

[25] J. Llorca, A. M. Tulino, K. Guan, J. Esteban, M. Varvello, N. Choi, and D. C. Kilper, "Dynamic in-network caching for energy efficient content delivery," in *Proc. IEEE INFOCOM'13, Turin, Italy*, April 2013, pp. 245–249.

[26] B. Perabathini, E. Bastug, M. Kountouris, M. Debbah, and A. Conte, "Caching at the edge: A green perspective for 5g networks," in *Proc. IEEE ICC'15, London, UK*, June 2015, pp. 2830–2835.

[27] A. Y. S. Lam and V. O. K. Li, "Chemical-reaction-inspired metaheuristic for optimization," *IEEE Trans. Evol. Comput.*, vol. 14, no. 3, pp. 381–399, 2010.

[28] K. Samdanis, X. Costa-Perez, and V. Sciancalepore, "From network sharing to multi-tenancy: The 5g network slice broker," *IEEE Commun. Mag.*, vol. 54, no. 7, pp. 32–39, July 2016.

[29] A. Fischer, J. F. Botero, M. T. Beck, H. de Meer, and X. Hesselbach, "Virtual network embedding: A survey," *IEEE Communications Surveys Tutorials*, vol. 15, no. 4, pp. 1888–1906, Fourth 2013.

[30] W. K. Chai, D. He, I. Psaras, and G. Pavlou, "Cache less for more in information-centric networks (extended version)," *Computer Communications*, vol. 36, no. 7, pp. 758–770, 2013.

[31] K. Ohsugi, J. Takemasa, Y. Koizumi, T. Hasegawa, and I. Psaras, "Power consumption model of ndn-based multicore software router based on detailed protocol analysis," *IEEE Journal on Selected Areas in Communications*, vol. 34, no. 5, pp. 1631–1644, May 2016.

[32] T. Hasegawa, Y. Nakai, K. Ohsugi, J. Takemasa, Y. Koizumi, and I. Psaras, "Empirically modeling how a multicore software icn router and an icn network consume power," in *Proceedings of the 1st international conference on Information-centric networking*. ACM, 2014, pp. 157–166.

[33] Y. Xu, Y. Li, Z. Wang, T. Lin, G. Zhang, and S. Ci, "Coordinated caching model for minimizing energy consumption in radio access network," in *Proc. IEEE ICC'14*, June 2014, pp. 2406–2411.

[34] R. Liu, H. Yin, X. Cai, G. Zhu, L. Yu, Q. Fan, and J. Xu, "Cooperative caching scheme for content oriented networking," *IEEE Commun. Letters*, vol. 17, no. 4, pp. 781–784, April 2013.

[35] A. Khreishah and J. Chakareski, "Collaborative caching for multicell-coordinated systems," in *Proc. IEEE INFOCOM'15 WKSHPS*, April 2015, pp. 257–262.

[36] R. Xie, T. Huang, F. R. Yu, and Y. Liu, "Caching design in green content centric networking based on chemical reaction optimization," in *Proc. IEEE GreenCom'13, Beijing, China*, Aug 2013, pp. 46–50.

[37] A. Y. S. Lam and V. O. K. Li, "Chemical reaction optimization for cognitive radio spectrum allocation," in *Proc. IEEE GLOBECOM'10, Miami, Florida, USA*, Dec 2010, pp. 1–5.

[38] J. Xu, A. Y. S. Lam, and V. O. K. Li, "Chemical reaction optimization for task scheduling in grid computing," *IEEE Trans. on Parallel and Distributed Syst.*, vol. 22, no. 10, pp. 1624–1631, Oct 2011.

[39] Y. Wang, Z. Li, G. Tyson, S. Uhlig, and G. Xie, "Design and evaluation of the optimal cache allocation for content-centric networking," *IEEE Trans. Comput.*, vol. 65, no. 1, pp. 95–107, Jan 2016.

[40] H. Hsu and K. C. Chen, "A resource allocation perspective on caching to achieve low latency," *IEEE Commun. Letters*, vol. 20, no. 1, pp. 145–148, Jan 2016.

[41] A. Y. S. Lam, J. Xu, and V. O. K. Li, "Chemical reaction optimization for population transition in peer-to-peer live streaming," in *Proc. IEEE Congr. Evol. Comput.*, July 2010, pp. 1–8.

Chapter 6

Joint optimization of 3C

In this chapter, we jointly consider computation offloading, spectrum and computation resource allocation and content caching in order to improve the performance of wireless cellular networks with mobile edge computing.

6.1 Introduction

With the radically increasing popularity of smart phones, new mobile applications such as face recognition, natural language processing, augmented reality, etc. are emerging constantly. However, traditional wireless cellular networks are becoming incapable meeting the exponentially growing demand not only in *high data rate* but also in *high computational capability* [1].

In order to address the data rate issue, the heterogeneous network structure was recently proposed, in which multiple low-power, local coverage enhancing small cells are deployed in one macro cell [2]. Since the same radio resource could be shared among small cells and the macro cell, small cell networks have been considered a promising solution to improving spectrum efficiency and energy efficiency, therefore being one of the key components of next generation wireless cellular networks [3]. Nevertheless, severe inter-cell interference may be incurred due to spectrum reuse, which will significantly deteriorate network performance. Without an effective *spectrum resource allocation mechanism*, the overall spectrum efficiency and energy efficiency of the network might become even worse than that of a network without small cells [4]. To address the spectrum allocation issue, the work in [5] proposes a graph coloring method to assign physical resource blocks (PRBs) to UE. The study of [4] presents a spectrum allocation algorithm based on game theory, in which the PRB allocation can reach a Nash equilibrium of the game.

On the other hand, to address the computational capability issue, mobile cloud computing (MCC) systems have been proposed to enable mobile devices to utilize the powerful computing capability in the cloud [6, 7]. In order to further reduce latency and make the solution more economical, *fog computing* has been proposed to deploy computing resources closer to end devices [8, 9, 10]. A similar technique, called *mobile edge computing* (MEC), has attracted great interest in wireless cellular networks recently [11, 12, 13]. MEC enables the mobile user's equipment (UE) to perform *computation offloading* to send their computation tasks to the MEC server via wireless cellular networks. Then each UE is associated with a clone in the MEC server, which executes the computation tasks on behalf of that UE. A number of previous works have discussed the computation offloading problem [14, 15, 16, 17, 18], from latency reduction and energy saving, or QoS (Quality of Service) promoting perspectives.

In addition, the server in the MEC system can realize an in-network caching function [11], similar to the function provided by information-centric networking (ICN) [19, 1, 21], which is able to reduce replicate information transmissions. According to the study of [22], in-network caching has the capability of significantly improving the quality of Internet content transmissions (e.g., reducing latency and increasing throughput) by moving the content closer to users. A number of research efforts have been dedicated to content caching strategies. The caching strategies proposed in [23] and [24] are based on where routers are located in the topology, while [25] designs a strategy according to content popularity.

Although some outstanding works have been dedicated to studying computation offloading, resource allocation and content caching, these important aspects were generally considered separately in the existing works. However, as shown in the following, it is necessary to jointly address these issues to improve the performance of next generation wireless networks. Therefore, in this chapter, we jointly consider computation offloading, spectrum and computation resource allocation and content caching in order to improve the performance of wireless cellular networks with mobile edge computing. The motivations behind our work are based on the following observations.

■ Computation offloading, resource allocation and content caching are all parts of the entire system, and they all contribute to the end-to-end user experience, which cannot be guaranteed by the optimization of one single segment of the whole system [26].

■ If multiple UE choose to offload their computation tasks to the MEC server via small cell networks simultaneously, severe interference can be generated, which will decrease the data rate. Moreover, the MEC server could be overloaded. In this case, it is not beneficial for all the UE to offload tasks to the MEC server. Instead, some UE should be selected to offload computations, while others should execute computations locally.

∎ Different amounts of spectrum and computation resources should be allocated to different UE to fulfill different user demands.

∎ Due to the limited caching space of the MEC server, different caching strategies should be applied upon different contents, in order to maximize caching revenue.

Therefore, an integrated framework for computation offloading, resource allocation and content caching has the potential to significantly improve the performance of wireless cellular networks with mobile edge computing.

To the best of our knowledge, the joint design of computation offloading, resource allocation and content caching has not been addressed in previous works. The distinct features of this chapter are as follows.

∎ We formulate the computation offloading decision, resource allocation and content caching in wireless cellular networks with mobile edge computing as an optimization problem.

∎ We transform the original non-convex problem into a convex problem and provide proof of the convexity of the transformed problem.

∎ We decompose the problem and apply the alternating direction method of multipliers (ADMM) to solve the problem in an efficient and practical way.

∎ Simulation results are presented to show the effectiveness of the proposed scheme with different system parameters.

6.2 System model

In this section, the system model adopted in this work is described. We first describe the network model, then we present the communication model, computation model and caching model in detail. Finally, the utility function of the optimization problem is proposed.

6.2.1 Network model

An environment of one macro cell and N small cells in the terminology of LTE standards is considered here. The macro cell is connected to the Internet through the core network of a cellular communication system. A MEC server is placed in the macro eNodeB (MeNB), and all the N small cell eNodeBs (SeNBs) are connected to the MeNB as well as the MEC server. In this chapter, it is assumed that the SeNBs are connected to the MeNB in a wired manner

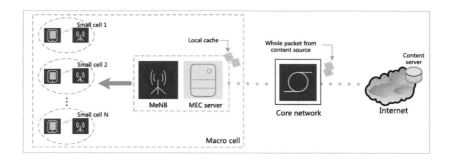

Figure 6.1: Network model.

[27]. The set of small cells is denoted by $\mathcal{N} = \{1, 2, ..., N\}$, and we use n to refer to the nth small cell (SeNB). It is assumed that SeNB n is associated with K_n mobile UE. We let $\mathcal{K}_n = \{1, 2, ..., K_n\}$ denote the set of UE associating with SeNB n, and k_n refers to the kth UE which associates with the nth SeNB. In this chapter, we consider single-antenna UE and SeNBs. The network model is illustrated in Figure 6.1.

We assume that each UE has a computationally intensive and delay sensitive task to be completed. Each UE can offload the computation to the MEC server through the SeNB with which it is associated, or execute the computation task locally. UE can request content from the Internet, then the Internet content will be transmitted through the macro base station (MeNB) to UE. Upon the first transmission of any particular Internet content, the MEC server can choose whether to store the content or not. If the content is stored, it can be used by other UE without another transmission from the Internet in the future. In this chapter, we consider two logical roles in the network: *mobile network operator* (MNO) and *MEC system operator* (MSO). The mobile network operators possess and operate the radio resources and physical infrastructure of the wireless network, including spectrum, backhaul, radio access networks, transmission networks, core networks, etc., while the MEC system operators own the MEC servers, lease physical resources (e.g., spectrum and backhaul) from the MNO and provide mobile edge computing services to UE. The MSO will charge the UE for receiving mobile edge computing services.

Similar to many previous works in mobile cloud computing [28] and mobile networking [29, 30, 31, 32, 33], to enable tractable analysis and get useful insights, we employ a quasi-static scenario where the set of mobile device users $\mathcal{K}_n, \forall n$ remains unchanged during a computation offloading period (e.g., within several seconds), while it may change across different periods. Since both the communication and computation aspects play a key role in mobile edge computing, next the communication and computation models are introduced in detail.

The notations that will be used in the rest of this chapter are summarized in Table 6.1.

Table 6.1: Notation

Sym.	Definition	Sym.	Definition
$\mathcal{N}$	The set of small cells.	$\mathbf{s}$	Radio spectrum allocation vector.
$\mathcal{K}_n$	The set of UE associated with SeNB n.	$\mathbf{c}$	Computation resource allocation profile.
k_n	kth UE associating with the nth SeNB.	$\mathbf{h}$	Internet content caching decision vector.
p_{k_n}	The transmission power of UE k_n.	δ_n	Unit price for leasing spectrum from small cell n.
N	Total number of small cells.	η_n	Unit price for leasing backhaul from small cell n.
B	Total system bandwidth.	θ_n	Unit price for transmitting computation input data.
F	Total computational capability of the MEC server.	λ	Unit price for leasing computation resource.
L	Backhaul capacity between MeNB and MEC server.	ζ	Unit price for leasing backhaul connecting Internet.
L_n	Backhaul capacity of SeNB n.	ϖ	The cost in the memory for caching one content.
Y	Total storage capability of the MEC server.	ι_{k_n}	Revenue of assigning radio resource to UE k_n.
$G_{k_n,n}$	The channel gain between UE k_n and SeNB n.	Ω_{k_n}	Revenue of allocating comput. resource to UE k_n.
W_{k_n}	Computation task of UE k_n.	Λ_{k_n}	Revenue of caching the content of UE k_n.
$f^{(l)}_{k_n}$	Computation capability of UE k_n.	$f^{(e)}_{k_n}$	Computation resource assigned to UE k_n.
$\mathbf{a}$	Computation offloading decision vector.		

6.2.2 Communication model

Every SeNB in the network is linked to the MEC server, so each UE could offload its computation task to the MEC server via the SeNB to which it is connected. We denote $a_{k_n} \in \{0, 1\}, \forall n, k$ as the computation offloading decision of UE k_n. Specifically, we have $a_{k_n} = 0$ if UE k_n was determined to compute its task locally on the mobile device. We have $a_{k_n} = 1$ if UE k_n was chosen to offload the computation to the MEC server via wireless access. So we have $\boldsymbol{a} = \{a_{k_n}\}_{k_n \in \mathcal{K}_n, n \in \mathcal{N}}$ as the offloading decision profile.

In this chapter, we consider the case where the spectrum used by small cells is overlaid, which means there exists interference between small cells. However, the spectrum within one small cell is orthogonally assigned to every UE, so there will be no interference within one small cell. Only uplink direction transmissions are considered, which means transmission is from a UE to the SeNB to which it is associated, and interference is from a UE to a neighboring SeNB. In this chapter, we assume that the interference only occurs when UE served by various SeNBs occupy the same frequency simultaneously. The whole available spectrum bandwidth is B Hz. The backhaul capacity between the MeNB and the MEC server is L bps, and the backhaul capacity of SeNB n is L_n bps. According to the Shannon bound, the spectrum efficiency of UE k_n is given by

$$
e_{k_n} = \log_2\left(1 + \frac{p_{k_n} G_{k_n, n}}{\sigma + \sum_{m=1, m\neq n}^{N} \sum_{i=1}^{K_m} p_{i_m} G_{i_m, n}}\right), \quad \forall n, k, \tag{6.1}
$$

where p_{k_n} is the transmission power density of UE k_n, and $G_{k_n,n}$, $G_{i_m,n}$ stand for the channel gain between UE k_n and SeNB n, and the channel gain between UE i_m and SeNB n, respectively. σ denotes the power spectrum density of additive white Gaussian noise.

We denote $s_{k_n} \in [0, 1], \forall n, k$ as the percentage of radio spectrum allocated to UE k_n by small cell n; thus $\sum_{k_n \in \mathcal{K}_n} s_{k_n} \leq 1, \forall n$. We have $\boldsymbol{s} = \{s_{k_n}\}_{k_n \in \mathcal{K}_n, n \in \mathcal{N}}$ as the radio spectrum allocation profile. Then the expected instantaneous data rate of UE k_n, R_{k_n} is calculated as

$$
R_{k_n}(\boldsymbol{a}, \boldsymbol{s}) = a_{k_n} s_{k_n} B e_{k_n}, \quad \forall n, k. \tag{6.2}
$$

The data rate cannot exceed the backhaul capacity of SeNB n; thus $\sum_{k_n \in \mathcal{K}_n} R_{k_n} \leq L_n, \forall n$ must hold. The total data rate of all the UE cannot exceed the backhaul capacity of the MeNB; thus $\sum_{n \in \mathcal{N}} \sum_{k_n \in \mathcal{K}_n} R_{k_n} \leq L$ must hold.

6.2.3 Computation model

For the computation model, we consider that each UE k_n has a computation task $W_{k_n} \triangleq (Z_{k_n}, D_{k_n})$, which can be computed either locally on the mobile device or remotely on the MEC server via computation offloading, as in [16]. Here Z_{k_n} stands for the size of input data, including program codes and input parameters, and D_{k_n} denotes the total number of CPU cycles required to accomplish the computation task W_{k_n}. A UE k_n can use the method in [28] and [34] to obtain the information of Z_{k_n} and D_{k_n}. We next discuss the computation overhead in terms of processing time for both local and MEC computing approaches.

6.2.3.1   Local computing

For the local computing approach, the computation task W_{k_n} is executed locally on each mobile device. We denote $f_{k_n}^{(l)}$ as the computational capability (i.e., CPU cycles per second) of UE k_n. It is allowed that different UE may have different computational capabilities. The computation execution time $T_{k_n}^{(l)}$ of task W_{k_n} executed locally by UE k_n is expressed as

$$T_{k_n}^{(l)} = \frac{D_{k_n}}{f_{k_n}^{(l)}}. \tag{6.3}$$

6.2.3.2   MEC server computing

For the MEC server computing approach, a UE k_n will offload its computation task W_{k_n} through wireless access to ScNB n, then through the connection from SeNB n to the MEC server. Then the MEC server will execute the computation task instead of UE k_n. For offloading the computation task, a UE k_n will incur the consumption on time when transmitting computation input data to the MEC server. According to the communication model presented in Subsection 6.2.2, the time costs for transmitting the computation input data of size Z_{k_n} are calculated as

$$T_{k_n,off}^{(e)}(\boldsymbol{a}, \boldsymbol{s}) = \frac{Z_{k_n}}{R_{k_n}(\boldsymbol{a}, \boldsymbol{s})}. \tag{6.4}$$

The MEC server will execute the computation task after offloading. Let $f_{k_n}^{(e)}$ denote the computational capability (i.e., CPU cycles per second) of the MEC server assigned to UE k_n. Then the execution time of the MEC server on task W_{k_n} is given as

$$T_{k_n,exe}^{(e)} = \frac{D_{k_n}}{f_{k_n}^{(e)}}. \tag{6.5}$$

Then the total execution time of the task of UE k_n is given by

$$T_{k_n}^{(e)}(\boldsymbol{a}, \boldsymbol{s}) = T_{k_n,off}^{(e)}(\boldsymbol{a}, \boldsymbol{s}) + T_{k_n,exe}^{(e)}. \tag{6.6}$$

In Section 6.5, we will use this expression to assess the average UE time consumption for executing computation tasks in simulations.

Similar to the study in [16], the time consumption of computation outcome transmission from the MEC server to UE k_n is neglected in this work, due to the fact that the size of computation outcome data in general is much smaller than that of the computation input data including mobile system settings, program codes and input parameters.

6.2.4 Caching model

We denote $h_{k_n} \in \{0,1\}, \forall n, k$ as the caching strategy for UE k_n. Specifically, we have $h_{k_n} = 1$ if the MEC server decides to cache the content requested by UE k_n and $h_{k_n} = 0$ otherwise. So we have $\boldsymbol{h} = \{h_{k_n}\}_{k_n \in \mathcal{K}_n, n \in \mathcal{N}}$ as the caching decision profile.

According to [35] and [36], the reward of caching in wireless networks can be the reduction of backhaul delay or the alleviation of backhaul bandwidth. In this chapter the alleviated backhaul bandwidth between the macro cell and the Internet is adopted as the caching reward. Thus, the reward (alleviated backhaul bandwidth) of caching the content requested by UE k_n can be given as

$$Caching\ reward = q_{k_n} \bar{R} h_{k_n}, \tag{6.7}$$

where $\bar{R}$ is the average single UE data rate in the system, and q_{k_n} is the request rate (by other UE) of the content first requested by UE k_n. According to the statistics of [37], if the requested content has a constant size, the request rate follows a Zipf popularity distribution and therefore can be calculated as $q(i) = 1/i^\beta$, where i stands for the i-th most popular content, and β is a constant whose typical value is 0.56 [38]. Therefore, if the size of the content first requested by UE k_n is known, the request rate of other UE upon the same content could be derived from the equation given above. In fact, the modeling of request rates of the caching content is still under research by many scholars. Since we adopt constant request rates in this chapter, the modeling of request rates is outside the scope of this chapter.

It is worth noting that the storage capability of the MEC server is not unlimited; thus the sum size of all the cached content cannot exceed the total storage capability of the MEC server. In other words, $\sum_{n \in \mathcal{N}} \sum_{k_n \in \mathcal{K}_n} h_{k_n} o_{k_n} \leq Y$ must hold, where Y is the total storage capability of the MEC server, and o_{k_n} is the size of the content first requested by UE k_n. In this chapter, it is assumed that $o_{k_n} \forall k, n$ is constant and we adopt $o_{k_n} = 1$.

6.2.5 Utility function

In this chapter, we set the maximization of the revenue of the MSO as our goal. The MSO rents spectrum and backhaul from the MNO, and the unit price for leasing spectrum from small cell n is defined as δ_n per Hz, while the unit price of backhaul between small cell n and the macro cell is defined as η_n per bps. The MSO will charge UE for transmitting computation input data to the MEC server, and the unit price being charged is defined as θ_n per bps. So the net revenue of the MSO for assigning radio resources to UE k_n is calculated as $\iota_{k_n} = s_{k_n} \Psi_{k_n} = s_{k_n}(\theta_n Be_{k_n} - \delta_n B - \eta_n Be_{k_n})$.

We next calculate the revenue of MSO for allocating computation resources to UE. First, we define $c_{k_n} \in [0,1], \forall n, k$ as the percentage of the MEC server computation resource allocated to UE k_n; thus $\sum_{n \in \mathcal{N}} \sum_{k_n \in \mathcal{K}_n} c_{k_n} \leq 1$. We have $c = \{c_{k_n}\}_{k_n \in \mathcal{K}_n, n \in \mathcal{N}}$ as the computation resource allocation profile. Without losing generality, the different sizes of computation tasks and different local computation capability of different UE should be taken into account. So we assume that the MSO will charge UE k_n only for the difference between the MEC computation resource allocated to every unit computation task and the local computation resource assigned to every unit computation task, and the unit price is λ_n for small cell n. Then the net revenue of allocating computation resources to UE k_n is given as $\Omega_{k_n} = \lambda_n(c_{k_n} F/D_{k_n} - f_{k_n}^{(l)}/D_{k_n})$, where F stands for the total computation resource of the MEC server. Note that the reciprocal of $c_{k_n} F/D_{k_n}$ is the time consumption for the MEC server executing computation task D_{k_n}, and the reciprocal of $f_{k_n}^{(l)}/D_{k_n}$ is the time consumption for UE k_n locally executing task D_{k_n}. This implies that the amount of computation resources assigned to every unit computation task can reflect the time consumption of executing this task.

We next discuss the revenue of the MSO for caching Internet content requested by UE. We define the unit price of leasing backhaul between the macro cell and the Internet as ζ per bps, and the cost in the memory for caching one content is ϖ. If the content first requested by UE k_n was stored by the MEC server, the alleviated backhaul bandwidth in the future should be $\zeta q_{k_n} \bar{R}$. And the memory cost for storing that content is ϖ. So the long term revenue of caching the Internet content first requested by UE k_n is calculated as $\Lambda_{k_n} = \zeta q_{k_n} \bar{R} - \varpi$.

Next we formulate the utility function of the MSO as

$$
\begin{aligned}
U &= \sum_{n \in \mathcal{N}} \sum_{k_n \in \mathcal{K}_n} u\left(a_{k_n} \iota_{k_n} + a_{k_n} \Omega_{k_n}\right) + h_{k_n} \Lambda_{k_n} \\
&= \sum_{n \in \mathcal{N}} \sum_{k_n \in \mathcal{K}_n} u\left[a_{k_n} s_{k_n} \Psi_{k_n} + a_{k_n} \lambda_n \left(\frac{c_{k_n} F}{D_{k_n}} - \frac{f_{k_n}^{(l)}}{D_{k_n}}\right)\right] \\
&\quad + h_{k_n} \Lambda_{k_n} \\
&= \sum_{n \in \mathcal{N}} \sum_{k_n \in \mathcal{K}_n} a_{k_n} u\left(s_{k_n} \Psi_{k_n} + c_{k_n} \frac{\lambda_n F}{D_{k_n}} - \frac{\lambda_n f_{k_n}^{(l)}}{D_{k_n}}\right) \\
&\quad + h_{k_n} \Lambda_{k_n},
\end{aligned}
\tag{6.8}
$$

where $u(\cdot)$ is a utility function which is nondecreasing and convex. Since $h_{k_n} \Lambda_{k_n}$ is always non-negative due to problem optimality, it can be put outside the function $u(\cdot)$. It is equivalent to take a_{k_n} outside the function $u(\cdot)$. If $a_{k_n} = 0$, it means UE k_n will not offload the task to the MEC server, so the MSO will not earn, then $a_{k_n} u(s_{k_n}, c_{k_n}) = u(a_{k_n}, s_{k_n}, c_{k_n}) = 0$; if $a_{k_n} = 1$, it means the MSO may earn, and $a_{k_n} u(s_{k_n}, c_{k_n}) = u(a_{k_n}, s_{k_n}, c_{k_n})$. Here the logarithmic function, which has been used frequently in the literature [39], is adopted as the utility function, given as $u(x) = logx$ when $x > 0$ and $u(x) = -\infty$ otherwise.

Define

$$U' = \sum_{n \in \mathcal{N}} \sum_{k_n \in \mathcal{K}_n} a_{k_n} u \left(s_{k_n} \Psi_{k_n} + c_{k_n} \frac{\lambda_n F}{D_{k_n}} \right) + h_{k_n} \Lambda_{k_n}. \tag{6.9}$$

Because $\lambda_n f_{k_n}^{(l)}/D_{k_n}$ is constant, when U' reaches the maximum value, U reaches the maximum as well, i.e., the MSO reaches the maximum income. Let $\lambda_n F/D_{k_n} = \Phi_{k_n}$. Next we will use

$$U' = \sum_{n \in \mathcal{N}} \sum_{k_n \in \mathcal{K}_n} a_{k_n} u \left(s_{k_n} \Psi_{k_n} + c_{k_n} \Phi_{k_n} \right) + h_{k_n} \Lambda_{k_n} \tag{6.10}$$

as our objective function of the optimization problem.

6.3 Problem formulation, transformation and decomposition

In order to maximize the utility function of the MSO, we formulate it as an optimization problem and transform it into a convex optimization problem.

6.3.1 Problem formulation

We adopt the utility function proposed in (6.10) as the objective function of our optimization problem, and the problem is formulated as

$$\underset{a,s,c,h}{\text{Maximize}} \quad \sum_{n \in \mathcal{N}} \sum_{k_n \in \mathcal{K}_n} a_{k_n} u \left(s_{k_n} \Psi_{k_n} + c_{k_n} \Phi_{k_n} \right) + h_{k_n} \Lambda_{k_n}$$

$$s.t. \quad C1: \quad \sum_{k_n \in \mathcal{K}_n} a_{k_n} s_{k_n} \leq 1, \forall n$$

$$C2: \quad \sum_{k_n \in \mathcal{K}_n} a_{k_n} s_{k_n} Be_{k_n} \leq L_n, \forall n$$

$$C3: \quad \sum_{m \in \mathcal{N}/\{n\}} \sum_{k_m \in \mathcal{K}_m} a_{k_m} p_{k_m} G_{k_m,n} \leq I_n, \forall n \tag{6.11}$$

$$C4: \quad \sum_{n \in \mathcal{N}} \sum_{k_n \in \mathcal{K}_n} a_{k_n} c_{k_n} \leq 1$$

$$C5: \quad a_{k_n} \left(\frac{c_{k_n} F}{D_{k_n}} - \frac{f_{k_n}^{(l)}}{D_{k_n}} \right) \geq 0, \forall k, n$$

$$C6: \quad \sum_{n \in \mathcal{N}} \sum_{k_n \in \mathcal{K}_n} h_{k_n} \leq Y.$$

The first set of constraints (6.11) $C1$ guarantees that in every small cell, the sum of spectrum allocated to all the offloading UE cannot exceed the total available spectrum of that small cell. Constraints (6.11) $C2$ mean the sum data rate of all offloading UE which associate with SeNB n cannot exceed the backhaul capacity of small cell n. If too many UE are allowed to offload computation tasks to the MEC server, the transmitting delay will be high due to high interference. In order to guarantee a relatively high data rate, constraints (6.11) $C3$ are proposed to ensure that the interference on SeNB n caused by all offloading UE which are served by other SeNBs doesn't exceed a predefined threshold, I_n. Constraint (6.11) $C4$ is due to the request that the sum of computation resources allocated to all offloading UE in the whole system cannot exceed the total amount of computation resources (total computational capability) of the MEC server. Because we removed $(-\lambda_n f_{k_n}^{(l)}/D_{k_n})$ in (6.8), we need constraints (6.11) $C5$ to guarantee that the computation resources allocated to each offloading UE k_n is no less than that of itself. Constraint (6.11) $C6$ guarantees that the sum of all the cached content doesn't exceed the total storage capability of the MEC server.

6.3.2 Problem transformation

Problem (6.11) is difficult to solve due to the following observations:

∎ Due to the fact that a and h are binary variables, the feasible set of problem (6.11) is not convex.

∎ There exist product relationships between $\{a_{k_n}\}$ and linear function of $\{s_{k_n}\}$, as well as $\{c_{k_n}\}$, so that the objective function of problem (6.11) is not a convex function.

∎ The problem is quite large. If we assume that the average number of UEs in one small cell is k, the number of variables in this problem could reach $4kN$, and the complexity for a central algorithm to find a globally optimal solution will be $\mathcal{O}((kN)^x)$ ($x > 0$, $x = 1$ implies a linear algorithm while $x > 1$ implies a polynomial time algorithm) even if we simply consider all the variables as binary variables. In addition, the number of small cells in one macro cell is increasing as time goes on, which results in an even more radically increasing complexity in our problem.

As shown, problem (6.11) is a mixed discrete and non-convex optimization problem, and such problems are usually considered NP-hard problems [40]. Therefore, a transformation and simplification of the original problem are necessary. The transformation of the problem is composed of the following two steps:

6.3.2.1   *Binary variable relaxation*

In order to transform the non-convex feasible set of problem (6.11) into a convex set, we need to relax binary variables $\boldsymbol{a}$ and $\boldsymbol{h}$ into real value variables as $0 \leq a_{k_n} \leq 1$, $0 \leq h_{k_n} \leq 1$ [40]. The relaxed variables can be interpreted as the time fraction of access to the MEC computation resources of UE k_n and the time fraction of sharing the content cache introduced by the request of UE k_n, respectively.

6.3.2.2   *Substitution of the product term*

Due to the non-convex objective function, the problem is still intractable even though we relax the variables. Next we propose a proposition of the equivalent problem of (6.11) to make the problem solvable.

Proposition 1: If we define $\tilde{s}_{k_n} = s_{k_n} a_{k_n}$, $\tilde{c}_{k_n} = c_{k_n} a_{k_n}$, and $a_{k_n} u[(\tilde{s}_{k_n} \Psi_{k_n} + \tilde{c}_{k_n} \Phi_{k_n})/a_{k_n}] = 0$ when $a_{k_n} = 0$, the following formulation (6.12) is equivalent to problem (6.11):

$$
\begin{aligned}
\underset{\boldsymbol{a}, \tilde{\boldsymbol{s}}, \tilde{\boldsymbol{c}}, \boldsymbol{h}}{\text{Maximize}} \quad & \sum_{n \in \mathcal{N}} \sum_{k_n \in \mathcal{K}_n} a_{k_n} u \left(\frac{\tilde{s}_{k_n} \Psi_{k_n} + \tilde{c}_{k_n} \Phi_{k_n}}{a_{k_n}} \right) + h_{k_n} \Lambda_{k_n} \\
\text{s.t.} \quad C1 : \quad & \sum_{k_n \in \mathcal{K}_n} \tilde{s}_{k_n} \leq 1, \forall n \\
C2 : \quad & \sum_{k_n \in \mathcal{K}_n} \tilde{s}_{k_n} Be_{k_n} \leq L_n, \forall n \\
C3 : \quad & \sum_{m \in \mathcal{N}/\{n\}} \sum_{k_m \in \mathcal{K}_m} a_{k_m} p_{k_m} G_{k_m, n} < I_n, \forall n \\
C4 : \quad & \sum_{n \in \mathcal{N}} \sum_{k_n \in \mathcal{K}_n} \tilde{c}_{k_n} \leq 1 \\
C5 : \quad & \tilde{c}_{k_n} \frac{F}{D_{k_n}} - a_{k_n} \frac{f_{k_n}^{(l)}}{D_{k_n}} \geq 0, \forall k, n \\
C6 : \quad & \sum_{n \in \mathcal{N}} \sum_{k_n \in \mathcal{K}_n} h_{k_n} \leq Y \\
C7 : \quad & a_{k_n} \geq \tilde{s}_{k_n}, a_{k_n} \geq \tilde{c}_{k_n}, \forall k, n.
\end{aligned}
\tag{6.12}
$$

Proof: This proof of Proposition 1 is motivated by [41]. If we substitute $\tilde{s}_{k_n} = s_{k_n} a_{k_n}$ and $\tilde{c}_{k_n} = c_{k_n} a_{k_n}$ into (6.12), we can recover the original optimization problem (6.11) except the point when $a_{k_n} = 0$. Next we discuss this point. Suppose $a_{k_n} = 0$, then $s_{k_n} = 0$ and $c_{k_n} = 0$ will certainly hold because of the problem optimality. Apparently, if UE k_n will not offload computation tasks to the MEC server, SeNB n will not allocate any spectrum resource to UE n, and the MEC server will not assign any computation resources to it either. Thus, the complete mapping between $\{a_{k_n}, s_{k_n}, c_{k_n}\}$ and $\{a_{k_n}, \tilde{s}_{k_n}, \tilde{c}_{k_n}\}$ is as shown in (6.13) and (6.14).

$$
s_{k_n} = \begin{cases} \tilde{s}_{k_n}/a_{k_n}, & a_{k_n} > 0, \\ 0, & otherwise, \end{cases} \tag{6.13}
$$

$$
c_{k_n} = \begin{cases} \tilde{c}_{k_n}/a_{k_n}, & a_{k_n} > 0, \\ 0, & otherwise. \end{cases} \tag{6.14}
$$

Now it's a one-to-one mapping. Note that constraints (6.12) $C7$ guarantee that $\tilde{s}_{k_n}$ and $\tilde{c}_{k_n}$ don't exceed a_{k_n}, and that is because of $s_{k_n} \in [0,1]$ and $c_{k_n} \in [0,1]$. ∎

6.3.3 Convexity

In this subsection, we discuss the convexity of problem (6.12) using the well-known perspective function [42].

Proposition 2: If problem (6.12) is feasible, it is jointly convex with respect to all the optimization variables $\boldsymbol{a}$, $\tilde{\boldsymbol{s}}$, $\tilde{\boldsymbol{c}}$ and $\boldsymbol{h}$.

Proof: This proof of Proposition 2 is similar to [41]. $f(t,x) = x\log(t/x), t \geq 0, x \geq 0$ is the well-known perspective function of $f(x) = \log x$. Next we give a proof of the continuity of the perspective function $f(t,x) = x\log(t/x), t \geq 0, x \geq 0$ on the point $x = 0$. Let $s = t/x$,

$$
f(t,0) = \lim_{x \to 0} x\log\frac{t}{x} = \lim_{s \to \infty}\frac{t}{s}\log s = t\lim_{s \to \infty}\frac{\log s}{s} = 0. \tag{6.15}
$$

So we have $a_{k_n}\log[(\tilde{s}_{k_n}\Psi_{k_n} + \tilde{c}_{k_n}\Phi_{k_n})/a_{k_n}] = 0$ for $a_{k_n} = 0$. Since $(\tilde{s}_{k_n}\Psi_{k_n} + \tilde{c}_{k_n}\Phi_{k_n})$ are linear with respect to $\tilde{s}_{k_n}$ and $\tilde{c}_{k_n}$, $\log(\tilde{s}_{k_n}\Psi_{k_n} + \tilde{c}_{k_n}\Phi_{k_n})$ is a concave function. Then $a_{k_n}\log[(\tilde{s}_{k_n}\Psi_{k_n} + \tilde{c}_{k_n}\Phi_{k_n})/a_{k_n}]$ is concave due to the fact that it is the perspective function of $\log(\tilde{s}_{k_n}\Psi_{k_n} + \tilde{c}_{k_n}\Phi_{k_n})$. The perspective function of a concave function is concave [42]. Furthermore, $h_{k_n}\Lambda_{k_n}$ is linear; then it is obvious that our objective function of problem (6.12), i.e., $a_{k_n}\log[(\tilde{s}_{k_n}\Psi_{k_n} + \tilde{c}_{k_n}\Phi_{k_n})/a_{k_n}] + h_{k_n}\Lambda_{k_n}$ is concave. On the other hand, all the constraints of problem (6.12) are linear (the feasible set of the problem is a convex set), so problem (6.12) is a convex optimization problem. ∎

A lot of methods could be applied to solve a convex optimization problem. But as far as our problem (6.12) is concerned, as mentioned above, the problem becomes appreciably large as the number of small cells grows. In addition, if a centralized algorithm is adopted in the MEC server, the signaling overhead of delivering local information (e.g., channel status information (CSI)) to the

MEC server could be extremely high. Therefore, it will be more efficient to employ a distributed algorithm which runs on each SeNB as well as the MEC server. In the next section, we will decouple the optimization problem (6.12) in order to enable the application of a distributed optimization problem solving method, namely, the alternating direction method of multipliers (ADMM).

6.3.4 Problem decomposition

In order to make it possible for each SeNB to participate in the computation for problem solving, we need to separate problem (6.12) so that it can be solved in a distributed manner. However, optimization variables a, $\tilde{c}$ and h in problem (6.12) are considered global variables, which are not separable in the problem (specifically, it is constraints (6.12) $C3$, $C4$ and $C6$ that make the problem inseparable). Thus, in order to make the problem separable, we introduce local copies of the global variables. Since the global variables concern all the small cells in the network, which means they cannot be handled in any single small cell, we create a copy for every global variable in each small cell. Thus each small cell can independently conduct computation for problem solving with local copies. For small cell n, we denote $\hat{a}^n = \{\hat{a}^n_{k_j}\}_{k_j \in \mathcal{K}_j, j \in \mathcal{N}, n \in \mathcal{N}}$,
* $\hat{c}^n = \{\hat{c}^n_{k_j}\}_{k_j \in \mathcal{K}_j, j \in \mathcal{N}, n \in \mathcal{N}}$ and $\hat{h}^n = \{\hat{h}^n_{k_j}\}_{k_j \in \mathcal{K}_j, j \in \mathcal{N}, n \in \mathcal{N}}$ as the local copies of a, $\tilde{c}$ and h, respectively. We have

$$\begin{cases} \hat{a}^n_{k_j} = a_{k_j}, & \forall n, k, j, \\ \hat{c}^n_{k_j} = \tilde{c}_{k_j}, & \forall n, k, j, \\ \hat{h}^n_{k_j} = h_{k_j}, & \forall n, k, j. \end{cases} \quad (6.16)$$

Letting

$$U'' = \sum_{n \in \mathcal{N}} \sum_{k_n \in \mathcal{K}_n} \hat{a}^n_{k_n} u \left(\frac{\tilde{s}_{k_n} \Psi_{k_n} + \hat{c}^n_{k_n} \Phi_{k_n}}{\hat{a}^n_{k_n}} \right) + \hat{h}^n_{k_n} \Lambda_{k_n}, \quad (6.17)$$

---

*Here we need to introduce another small cell index $j \in \mathcal{N}$ to indicate each small cell in the local copy of small cell n.

next we give the equivalent global consensus version of problem (6.12) as

$$\underset{\substack{\{\hat{a}^n,\tilde{s},\hat{c}^n,\hat{h}^n\},\\ \{a,\tilde{c},h\}}}{\text{Maximize}}\ U''$$

$$
\begin{aligned}
s.t.\quad C1:\ & \sum_{k_n\in\mathcal{K}_n} \tilde{s}_{k_n} \leq 1, \forall n\\[4pt]
C2:\ & \sum_{k_n\in\mathcal{K}_n} \tilde{s}_{k_n} Be_{k_n} \leq L_n, \forall n\\[4pt]
C3:\ & \sum_{j\in\mathcal{N}/\{n\}}\sum_{k_j\in\mathcal{K}_j} \hat{a}_{k_j}^n p_{k_j} G_{k_j,n} \leq I_n, \forall n\\[4pt]
C4:\ & \sum_{j\in\mathcal{N}}\sum_{k_j\in\mathcal{K}_j} \hat{c}_{k_j}^n \leq 1, \forall n\\[4pt]
C5:\ & \hat{c}_{k_j}^n \frac{F}{D_{k_j}} - \hat{a}_{k_j}^n \frac{f_{k_j}^{(l)}}{D_{k_j}} \geq 0, \forall n,k,j\\[4pt]
C6:\ & \sum_{j\in\mathcal{N}}\sum_{k_j\in\mathcal{K}_j} \hat{h}_{k_j}^n \leq Y, \forall n\\[4pt]
C7:\ & \hat{a}_{k_n}^n \geq \tilde{s}_{k_n}, \hat{a}_{k_n}^n \geq \hat{c}_{k_n}^n, \forall k,n\\[4pt]
C8:\ & \hat{a}_{k_j}^n = a_{k_j}, \hat{c}_{k_j}^n = \tilde{c}_{k_j}, \hat{h}_{k_j}^n = h_{k_j}, \forall n,k,j.
\end{aligned}
\tag{6.18}
$$

The consensus constraint (6.18) $C8$ imposes that all the local copy variables in all small cells (i.e., $\{\hat{a}_{k_j}^n, \hat{c}_{k_j}^n, \hat{h}_{k_j}^n\}_{n\in\mathcal{N}}$) *must be consistent with* the corresponding global variables (i.e., $\{a_{k_j}, \tilde{c}_{k_j}, h_{k_j}\}$).

For ease of description, we define the following set as the local variable *feasible set* of each small cell $n\in\mathcal{N}$:

$$
\xi_n = \left\{
\begin{array}{c}
\hat{a}^n\\
\tilde{s}^n\\
\hat{c}^n\\
\hat{h}^n
\end{array}
\middle|
\begin{array}{l}
\sum_{k_n\in\mathcal{K}_n} \tilde{s}_{k_n} \leq 1\\
\sum_{k_n\in\mathcal{K}_n} \tilde{s}_{k_n} Be_{k_n} \leq L_n\\
\sum_{j\in\mathcal{N}/\{n\}}\sum_{k_j\in\mathcal{K}_j} \hat{a}_{k_j}^n p_{k_j} G_{k_j,n} \leq I_n\\
\sum_{j\in\mathcal{N}}\sum_{k_j\in\mathcal{K}_j} \hat{c}_{k_j}^n \leq 1\\
\hat{c}_{k_j}^n F/D_{k_j} - \hat{a}_{k_j}^n f_{k_j}^{(l)}/D_{k_j} \geq 0, \forall k,j\\
\sum_{j\in\mathcal{N}}\sum_{k_j\in\mathcal{K}_j} \hat{h}_{k_j}^n \leq Y\\
\hat{a}_{k_n}^n \geq \tilde{s}_{k_n}, \hat{a}_{k_n}^n \geq \hat{c}_{k_n}^n, \forall k
\end{array}
\right\}, \forall n.
\tag{6.19}
$$

Note that ξ_n is proprietary for small cell n and is completely decoupled from other small cells.

Next we give the local utility function of each small cell $n \in \mathcal{N}$ as follows:

$$v_n = \begin{cases} - \left[\sum_{k_n \in \mathcal{K}_n} \hat{a}_{k_n}^n u \left(\frac{\tilde{s}_{k_n} \Psi_{k_n} + \hat{c}_{k_n}^n \Phi_{k_n}}{\hat{a}_{k_n}^n} \right) + \hat{h}_{k_n}^n \Lambda_{k_n} \right], \\ \qquad when \quad \{\hat{a}^n, \tilde{s}_n, \hat{c}^n, \hat{h}^n\} \in \xi_n, \\ + \infty, \ otherwise. \end{cases} \quad (6.20)$$

With (6.19) and (6.20), an equivalent formulation of problem (6.18) is given as

$$\begin{aligned} \underset{\{\hat{a}^n, \tilde{s}, \hat{c}^n, \hat{h}^n\}, \{a, \tilde{c}, h\}}{\text{Minimize}} \quad & \sum_{n \in \mathcal{N}} v_n(\hat{a}^n, \tilde{s}_n, \hat{c}^n, \hat{h}^n) \\ s.t. \quad C1: \quad & \hat{a}_{k_j}^n = a_{k_j}, \forall n, k, j \\ C2: \quad & \hat{c}_{k_j}^n = \tilde{c}_{k_j}, \forall n, k, j \\ C3: \quad & \hat{h}_{k_j}^n = h_{k_j}, \forall n, k, j. \end{aligned} \quad (6.21)$$

Now it is obvious that in problem (6.21) the objective functions v_n with feasible sets ξ_n are separable with respect to all the small cells in the system. But the *consensus constraints* (6.21) $C1$–$C3$ remain coupled upon all the small cells. That is exactly what we want. The separation of the objective functions enables each small cell to independently handle the subproblem related to itself, while the persistence of coupling of the consensus constraints (6.21) $C1$–$C3$ guarantees the consistency of all the local copies with each other, as well as with the real global variables. In the next section we will apply the *Alternating Direction Method of Multipliers (ADMM)* to solve the problem in a distributed fashion.

6.4   Problem solving via ADMM

In this section, first we will derive the augmented Lagrangian with corresponding global consensus constraints and formulate the ADMM iteration steps [43, 44, 45]; second, the update methods for ADMM iterations are presented; third, the relaxed variables are recovered to binary variables; finally, the overall algorithm is summarized.

6.4.1   Augmented Lagrangian and ADMM sequential iterations

According to [43], problem (6.21) is called a *global consensus problem*, due to the fact that all the local variables are consistent (with the global variables).

According to [43], the augmented Lagrangian of problem (6.21) is given as

$$L_\rho(\{\hat{a}^n, \tilde{s}_n, \hat{c}^n, \hat{h}^n\}_{n \in \mathcal{N}}, \{a, \tilde{c}, h\}, \{\sigma^n, \omega^n, \tau^n\}_{n \in \mathcal{N}}) =$$

$$\sum_{n \in \mathcal{N}} v_n(\hat{a}^n, \tilde{s}_n, \hat{c}^n, \hat{h}^n) + \sum_{n \in \mathcal{N}} \sum_{j \in \mathcal{N}} \sum_{k_j \in \mathcal{K}_j} \sigma^n_{k_j} (\hat{a}^n_{k_j} - a_{k_j})$$

$$+ \sum_{n \in \mathcal{N}} \sum_{j \in \mathcal{N}} \sum_{k_j \in \mathcal{K}_j} \omega^n_{k_j} (\hat{c}^n_{k_j} - \tilde{c}_{k_j}) + \sum_{n \in \mathcal{N}} \sum_{j \in \mathcal{N}} \sum_{k_j \in \mathcal{K}_j} \tau^n_{k_j} (\hat{h}^n_{k_j} - h_{k_j}) \tag{6.22}$$

$$+ \frac{\rho}{2} \sum_{n \in \mathcal{N}} \sum_{j \in \mathcal{N}} \sum_{k_j \in \mathcal{K}_j} (\hat{a}^n_{k_j} - a_{k_j})^2 + \frac{\rho}{2} \sum_{n \in \mathcal{N}} \sum_{j \in \mathcal{N}} \sum_{k_j \in \mathcal{K}_j} (\hat{c}^n_{k_j} - \tilde{c}_{k_j})^2$$

$$+ \frac{\rho}{2} \sum_{n \in \mathcal{N}} \sum_{j \in \mathcal{N}} \sum_{k_j \in \mathcal{K}_j} (\hat{h}^n_{k_j} - h_{k_j})^2,$$

where $\sigma^n = \{\sigma^n_{k_j}\}_{n \in \mathcal{N}}$, $\omega^n = \{\omega^n_{k_j}\}_{n \in \mathcal{N}}$ and $\tau^n = \{\tau^n_{k_j}\}_{n \in \mathcal{N}}$ are the Lagrange multipliers with respect to (6.18) $C8$, and $\rho \in \mathbb{R}_{++}$ is the so called *penalty parameter*, which is a constant parameter intended for adjusting the convergence speed of ADMM [43]. Compared to the standard Lagrangian, the additional ρ-terms in the augmented Lagrangian (6.22) can improve the property of the iterative method [46]. Please note that for any feasible solution, the ρ-terms added in the augmented Lagrangian (6.22) are actually *equal to zero* [43].

With ADMM being applied to solving problem (6.21), the following sequential iterative optimization steps are presented as findings [43].

Local variables:

$$\{\hat{a}^n, \tilde{s}_n, \hat{c}^n, \hat{h}^n\}^{[t+1]}_{n \in \mathcal{N}} = $$

$$\underset{\{\hat{a}^n_{k_j}, \tilde{s}_{kn}, \hat{c}^n_{k_j}, \hat{h}^n_{k_j}\}}{\arg \min} \left\{ \begin{array}{l} v_n(\hat{a}^n, \tilde{s}_n, \hat{c}^n, \hat{h}^n) \\[4pt] + \sum_{j \in \mathcal{N}} \sum_{k_j \in \mathcal{K}_j} \sigma^{n[t]}_{k_j} \left(\hat{a}^n_{k_j} - a^{[t]}_{k_j} \right) \\[4pt] + \sum_{j \in \mathcal{N}} \sum_{k_j \in \mathcal{K}_j} \omega^{n[t]}_{k_j} \left(\hat{c}^n_{k_j} - \tilde{c}^{[t]}_{k_j} \right) \\[4pt] + \sum_{j \in \mathcal{N}} \sum_{k_j \in \mathcal{K}_j} \tau^{n[t]}_{k_j} \left(\hat{h}^n_{k_j} - h^{[t]}_{k_j} \right) \\[4pt] + \frac{\rho}{2} \sum_{j \in \mathcal{N}} \sum_{k_j \in \mathcal{K}_j} \left(\hat{a}^n_{k_j} - a^{[t]}_{k_j} \right)^2 \\[4pt] + \frac{\rho}{2} \sum_{j \in \mathcal{N}} \sum_{k_j \in \mathcal{K}_j} \left(\hat{c}^n_{k_j} - \tilde{c}^{[t]}_{k_j} \right)^2 \\[4pt] + \frac{\rho}{2} \sum_{j \in \mathcal{N}} \sum_{k_j \in \mathcal{K}_j} \left(\hat{h}^n_{k_j} - h^{[t]}_{k_j} \right)^2 \end{array} \right\} \tag{6.23}$$

Global variables:

$$\{a\}^{[t+1]} = $$

$$\underset{\{a_{k_j}\}}{\arg \min} \left\{ \begin{array}{l} \sum_{n \in \mathcal{N}} \sum_{j \in \mathcal{N}} \sum_{k_j \in \mathcal{K}_j} \sigma^{n[t]}_{k_j} \left(\hat{a}^{n[t+1]}_{k_j} - a_{k_j} \right) \\[4pt] + \frac{\rho}{2} \sum_{n \in \mathcal{N}} \sum_{j \in \mathcal{N}} \sum_{k_j \in \mathcal{K}_j} \left(\hat{a}^{n[t+1]}_{k_j} - a_{k_j} \right)^2 \end{array} \right\} \tag{6.24}$$

$$\{\tilde{c}\}^{[t+1]} =$$

$$\underset{\{\tilde{c}_{k_j}\}}{\arg\min} \left\{ \begin{array}{c} \sum\limits_{n \in \mathcal{N}} \sum\limits_{j \in \mathcal{N}} \sum\limits_{k_j \in \mathcal{K}_j} \omega_{k_j}^{n[t]} \left(\hat{c}_{k_j}^{n[t+1]} - \tilde{c}_{k_j} \right) \\ + \frac{\rho}{2} \sum\limits_{n \in \mathcal{N}} \sum\limits_{j \in \mathcal{N}} \sum\limits_{k_j \in \mathcal{K}_j} \left(\hat{c}_{k_j}^{n[t+1]} - \tilde{c}_{k_j} \right)^2 \end{array} \right\} \tag{6.25}$$

$$\{h\}^{[t+1]} =$$

$$\underset{\{h_{k_j}\}}{\arg\min} \left\{ \begin{array}{c} \sum\limits_{n \in \mathcal{N}} \sum\limits_{j \in \mathcal{N}} \sum\limits_{k_j \in \mathcal{K}_j} \tau_{k_j}^{n[t]} \left(\hat{h}_{k_j}^{n[t+1]} - h_{k_j} \right) \\ + \frac{\rho}{2} \sum\limits_{n \in \mathcal{N}} \sum\limits_{j \in \mathcal{N}} \sum\limits_{k_j \in \mathcal{K}_j} \left(\hat{h}_{k_j}^{n[t+1]} - h_{k_j} \right)^2 \end{array} \right\} \tag{6.26}$$

Lagrange multipliers:

$$\{\sigma^n\}_{n \in \mathcal{N}}^{[t+1]} = \sigma^{n[t]} + \rho(\hat{a}^{n[t+1]} - a^{[t+1]}) \tag{6.27}$$

$$\{\omega^n\}_{n \in \mathcal{N}}^{[t+1]} = \omega^{n[t]} + \rho(\hat{c}^{n[t+1]} - \tilde{c}^{[t+1]}) \tag{6.28}$$

$$\{\tau^n\}_{n \in \mathcal{N}}^{[t+1]} = \tau^{n[t]} + \rho(\hat{h}^{n[t+1]} - h^{[t+1]}), \tag{6.29}$$

where the superscript $[t]$ stands for the iteration index.

It is obvious that the iteration steps (6.23) concerning local variables are completely separable with respect to the small cell index n, and thus can be executed by each SeNB. The iteration steps (6.24)–(6.29) concerning global variables and Lagrange multipliers would be executed by the MEC server. In the following subsections, we will discuss methods for solving these iterations.

6.4.2 Local variables update

As described above, iteration (6.23) is decomposed into N subproblems, with each of them being solved by an SeNB. Thus, after eliminating the constant terms, it is equivalent for SeNB $n \in \mathcal{N}$ to solve the following optimization problem at iteration $[t+1]$:

$$\underset{\substack{\{\hat{a}_{k_j}^n, \tilde{s}_{k_n}, \\ \hat{c}_{k_j}^n, \hat{h}_{k_j}^n\}}}{\text{Minimize}} \quad v_n(\hat{a}^n, \tilde{s}_n, \hat{c}^n, \hat{h}^n)$$

$$+ \sum_{j \in \mathcal{N}} \sum_{k_j \in \mathcal{K}_j} \left[\sigma_{k_j}^{n[t]} \hat{a}_{k_j}^n + \frac{\rho}{2} \left(\hat{a}_{k_j}^n - a_{k_j}^{[t]} \right)^2 \right]$$

$$+ \sum_{j \in \mathcal{N}} \sum_{k_j \in \mathcal{K}_j} \left[\omega_{k_j}^{n[t]} \hat{c}_{k_j}^n + \frac{\rho}{2} \left(\hat{c}_{k_j}^n - \tilde{c}_{k_j}^{[t]} \right)^2 \right] \tag{6.30}$$

$$+ \sum_{j \in \mathcal{N}} \sum_{k_j \in \mathcal{K}_j} \left[\tau_{k_j}^{n[t]} \hat{h}_{k_j}^n + \frac{\rho}{2} \left(\hat{h}_{k_j}^n - h_{k_j}^{[t]} \right)^2 \right]$$

$$\text{s.t.} \quad \{\hat{a}^n, \tilde{s}_n, \hat{c}^n, \hat{h}^n\} \in \xi_n.$$

Algorithm 3: Primal-dual interior-point method for local variables updating

---

1: Initialization

Given $\{\hat{a}^n, \tilde{s}_n, \hat{c}^n, \hat{h}^n\} \in \xi_n$, $\varrho > 0$, $\varsigma > 1$, $\epsilon_{feas} > 0$, $\epsilon > 0$.

2: **Repeat**

a) Determine t. Set $t := \varsigma m / \hat{\eta}$;

b) Compute primal-dual search direction $\Delta_{y_{pd}}$;

c) Line search and update

Determine step length $s > 0$ and set $y := y + s\Delta_{y_{pd}}$.

Until $\|r_{pri}\|_2 \le \epsilon_{feas}$, $\|r_{dual}\|_2 \le \epsilon_{feas}$, and $\hat{\eta} \le \epsilon$.

Obviously, problem (6.30) is a convex problem due to its quadric objective function and convex feasible set. So here we employ the *primal-dual interior-point method* [42] to solve this problem, which is briefly described in Algorithm 1.

In Algorithm 3, $\hat{\eta}$ stands for the surrogate duality gap, and m denotes the number of constraints. Due to limited space, a detailed description of this method is omitted here, and readers could turn to [42] for more information, where the method is described in detail.

We only need to consider how to provide an initial feasible solution $\{a, \tilde{c}, h\}^{[0]}$ for Algorithm 3. In order to do so, let us consider an extreme case, where only one UE among all the UE in system is allowed to offload a computation task to the MEC server. And we designate this UE as UE $\bar{k}_{\bar{j}}$, so we set $a_{\bar{k}_{\bar{j}}}^{[0]} = 1$ and $a_{k_j}^{[0]} = 0, \forall k_j \ne \bar{k}_{\bar{j}}$. Naturally, when only one UE is offloading a computation task to the MEC server, all the computation resources would be allocated to this UE; thus $\tilde{c}_{\bar{k}_{\bar{j}}}^{[0]} = 1$ and $\tilde{c}_{k_j}^{[0]} = 0, \forall k_j \ne \bar{k}_{\bar{j}}$. And all the radio resources of small cell $\bar{j}$ will be assigned to UE $\bar{k}_{\bar{j}}$, so $s_{\bar{k}_{\bar{j}}} = 1$, while $s_{k_{\bar{j}}} = 0, \forall k \ne \bar{k}$. As far as all the other small cells are concerned, they will not allocate any spectrum resources to any of their associating UE, since all their UE will execute the computation tasks locally. Thus $s_{k_j} = 0, \forall k_j \ne \bar{k}_{\bar{j}}$. As far as the caching strategy is concerned, we assume that the MEC server chose to store the Internet content requested by UE $\bar{k}_{\bar{j}}$ and not to store content requested by any other UE; thus we have $h_{\bar{k}_{\bar{j}}}^{[0]} = 1$ and $h_{k_j}^{[0]} = 0, \forall k_j \ne \bar{k}_{\bar{j}}$. By doing so, the constraints of problem (6.30) are automatically satisfied.

6.4.3 *Global variables and Lagrange multipliers update*

Now we move on to the global variables. Since problems (6.24), (6.25) and (6.26) are unconstrained quadratic problems and are strictly convex due to the added quadratic regularization terms in the augmented Lagrangian (6.22), we can solve them by simply setting the *gradients* of $\boldsymbol{a}$, $\tilde{\boldsymbol{c}}$ and $\boldsymbol{h}$ to zero, i.e.,

$$\sum_{n \in \mathcal{N}} \sigma_{k_j}^{n[t]} + \rho \sum_{n \in \mathcal{N}} \left(\hat{a}_{k_j}^{n[t+1]} - a_{k_j} \right) = 0, \quad \forall k, j \tag{6.31}$$

$$\sum_{n \in \mathcal{N}} \omega_{k_j}^{n[t]} + \rho \sum_{n \in \mathcal{N}} \left(\hat{c}_{k_j}^{n[t+1]} - \tilde{c}_{k_j} \right) = 0, \quad \forall k, j \tag{6.32}$$

$$\sum_{n \in \mathcal{N}} \tau_{k_j}^{n[t]} + \rho \sum_{n \in \mathcal{N}} \left(\hat{h}_{k_j}^{n[t+1]} - h_{k_j} \right) = 0, \quad \forall k, j \tag{6.33}$$

and this results in

$$a_{k_j}^{[t+1]} = \frac{1}{N\rho} \sum_{n \in \mathcal{N}} \sigma_{k_j}^{n[t]} + \frac{1}{N} \sum_{n \in \mathcal{N}} \hat{a}_{k_j}^{n[t+1]}, \quad \forall k, j \tag{6.34}$$

$$\tilde{c}_{k_j}^{[t+1]} = \frac{1}{N\rho} \sum_{n \in \mathcal{N}} \omega_{k_j}^{n[t]} + \frac{1}{N} \sum_{n \in \mathcal{N}} \hat{c}_{k_j}^{n[t+1]}, \quad \forall k, j \tag{6.35}$$

$$h_{k_j}^{[t+1]} = \frac{1}{N\rho} \sum_{n \in \mathcal{N}} \tau_{k_j}^{n[t]} + \frac{1}{N} \sum_{n \in \mathcal{N}} \hat{h}_{k_j}^{n[t+1]}, \quad \forall k, j. \tag{6.36}$$

By initializing the Lagrange multipliers as zeros at iteration $[t]$ [43], i.e., $\sum_{n \in \mathcal{N}} \sigma_{k_j}^{n[t]} = 0$, $\sum_{n \in \mathcal{N}} \omega_{k_j}^{n[t]} = 0$, $\sum_{n \in \mathcal{N}} \tau_{k_j}^{n[t]} = 0$, $\forall k, j$, equations (6.34)–(6.36) reduce to

$$a_{k_j}^{[t+1]} = \frac{1}{N} \sum_{n \in \mathcal{N}} \hat{a}_{k_j}^{n[t+1]}, \quad \forall k, j \tag{6.37}$$

$$\tilde{c}_{k_j}^{[t+1]} = \frac{1}{N} \sum_{n \in \mathcal{N}} \hat{c}_{k_j}^{n[t+1]}, \quad \forall k, j \tag{6.38}$$

$$h_{k_j}^{[t+1]} = \frac{1}{N} \sum_{n \in \mathcal{N}} \hat{h}_{k_j}^{n[t+1]}, \quad \forall k, j. \tag{6.39}$$

Equations (6.37)–(6.39) imply that at each iteration the global variables are calculated by averaging out all the corresponding local copies in all the small cells, which can be philosophically interpreted as the summary of the small cells' opinions on the optimal global variables.

The process of Lagrange multipliers $\{\boldsymbol{\sigma}^n, \boldsymbol{\omega}^n, \boldsymbol{\tau}^n\}_{n \in \mathcal{N}}$ updating is simple compared to $\{\hat{\boldsymbol{a}}^n, \tilde{\boldsymbol{s}}_n, \hat{\boldsymbol{c}}^n, \hat{\boldsymbol{h}}^n\}_{n \in \mathcal{N}}$ and $\{\boldsymbol{a}, \tilde{\boldsymbol{c}}, \boldsymbol{h}\}$ updating. With the current local variables received from each SeNB, the MEC server can easily obtain the Lagrange multipliers using equations (6.27)–(6.29) in each iteration.

6.4.4 Algorithm stopping criterion and convergence

Apparently, all the variables of problem (6.21) are bounded and the objective function of the problem is bounded, too, so inequality $\sum_{n \in \mathcal{N}} v_n(\hat{a}^{n*}, \tilde{s}_n^*, \hat{c}^{n*}, \hat{h}^{n*}) < \infty$ holds, where $\{\hat{a}^{n*}, \tilde{s}_n^*, \hat{c}^{n*}, \hat{h}^{n*}\}$ is the optimal solution of problem (6.21). Since problem (6.21) is a convex optimization problem (the proof of the convexity of the problem has been given in Section 6.3.3), strong duality holds [42]. According to [43], the objective function of problem (6.21) is convex, closed and proper, and the Lagrangian (6.22) has a saddle point, so the ADMM iterations described above satisfy residual convergence, objective convergence and dual variable convergence when $t \to \infty$.

For implementation purposes, we can employ the rational stopping criterion proposed in [43], which is given as

$$\| r_p^{[t+1]} \|_2 \leq \vartheta_{pri} \quad and \quad \| r_d^{[t+1]} \|_2 \leq \vartheta_{dual}, \tag{6.40}$$

where $\vartheta_{pri} > 0$ and $\vartheta_{dual} > 0$ are small positive constant scalars, which are called the feasibility tolerances for the primal and dual feasibility conditions, respectively. This stopping criterion implies that the primal residual $r_p^{[t+1]}$ and the dual residual $r_d^{[t+1]}$ must be small.

As this stopping criterion in [43] being applied to our algorithm, the residual for the primal feasibility condition of small cell n in iteration $[t+1]$ must be small enough so that

$$\| \hat{a}^{n[t+1]} - a^{[t+1]} \|_2 \leq \vartheta_{pri}, \quad \forall n, \tag{6.41}$$

$$\| \hat{c}^{n[t+1]} - \tilde{c}^{[t+1]} \|_2 \leq \vartheta_{pri}, \quad \forall n, \tag{6.42}$$

$$\| \hat{h}^{n[t+1]} - h^{[t+1]} \|_2 \leq \vartheta_{pri}, \quad \forall n, \tag{6.43}$$

and the residual for the dual feasibility condition in iteration $[t+1]$ must be small enough so that

$$\| a^{[t+1]} - a^{[t]} \|_2 \leq \vartheta_{dual}, \tag{6.44}$$

$$\| \tilde{c}^{[t+1]} - \tilde{c}^{[t]} \|_2 \leq \vartheta_{dual}, \tag{6.45}$$

$$\| h^{[t+1]} - h^{[t]} \|_2 \leq \vartheta_{dual}. \tag{6.46}$$

6.4.5 Binary variables recovery

In order to transform the original problem (6.11) into a convex problem, we have relaxed the binary variables a and h into continuous variables in Section 6.3.2.1. So we need to recover the binary variables after the convergence of the ADMM process. In order to maximize the revenue of the MSO, we try to maximize the number of offloading UE and to store as much Internet content as possible. The recovery deals with the marginal benefit of each UE. We adopt the following Algorithm 4 to recover the binary variables a and h. In Algorithm 4 we use a as an example, and the same algorithm is applied

---

Algorithm 4: Binary variables recovery

1: Computing first partial derivations
   Compute the first partial derivations of augmented Lagrangian
   $Q_{k_n} = \partial L_\rho / \partial a_{k_n}$ with respect to each a_{k_n}.
2: Sort all the partial derivations $Q_{k_n}, \forall k, n$ from largest to smallest.
   Mark them with $Q_1, Q_2...Q_i...$, and mark the corresponding a_{k_n} as
   $a_1, a_2...a_i...$
3: **For** i=1,2,..., **Do**
   Set $a_i = 1$ and $a_{i+1}, a_{i+2}, a_{i+3}... = 0$;
   If Any of the constrains (6.11) C1-C6 does not hold, **Then Break.**
   End for
4: Output the recovered binary variables $\{a_{k_n}\}, \forall k, n$.

---

to $\boldsymbol{h}$. It should be mentioned that the recovery of binary variables creates a gap between our results and the upper bound results. (Since the problem is NP-hard, it is very difficult to examine the existence of the optimal results.) However, as will be shown in the simulation, the gap is not significant.

6.4.6   Feasibility, complexity and summary of the algorithm

If the computational capability and storage capability of the MEC server are too low, or the cost of the spectrum is too high, the utility function of our problem may become non-positive under all possible solutions. In that case, the optimal solution would be $\{\boldsymbol{a}^*, \boldsymbol{s}^*, \boldsymbol{c}^*, \boldsymbol{h}^*\} = \{\overrightarrow{0}, \overrightarrow{0}, \overrightarrow{0}, \overrightarrow{0}\}$, which means that all the UE in system will execute their computation tasks locally, and no spectrum and computational resources are allocated to any UE. Besides, no requested Internet content will be stored by the MEC server. This case may be treated as the problem becomes infeasible. However, except for the extreme cases, the MEC server could allow at least one UE to offload its computation task or store at least one content, and gain a positive revenue, due to the fact that a typical MEC server could always have a much higher computational capability and storage capability than a single UE.

Now let us discuss the complexity of our algorithm by comparing it with the complexity of the centralized algorithm. First we assume that the average number of UE in each small cell is k, and the total number of small cells is N. Thus the size of input for the centralized algorithm would be kN. If the centralized algorithm adopted the primal-dual interior-point method for convex optimization problem solving at each input, the complexity would be $\mathcal{O}((kN)^x)$ with $x > 0$, where $x = 1$ implies a linear algorithm while $x > 1$ implies a polynomial time algorithm. Now we move on to our proposed distributed algorithm. In local variables updating (6.23), the size of input is k, due to the fact that each small cell only needs to mind its own associating

---

Algorithm 5: Decentralized resource allocation algorithm in MEC system via ADMM

1: Initialization

   a) MEC server determines the stopping criterion threshold ϑ_{pri} and ϑ_{dual};

   b) MEC server initializes the initial feasible solution $\{a, \tilde{c}, h\}^{[0]}$ as described in Section 6.4.2 and sends it to each SeNB;

   c) Each SeNB n collects CSI of all its associating UE;

   d) Each SeNB n determines its initial Lagrange multipliers vectors $\{\sigma^{n[0]} > 0, \omega^{n[0]} > 0, \tau^{n[0]} > 0\}$ and sends them to MEC server; $t=0$.

2: Iterations

   Repeat

   a) Each SeNB n updates its local variables $\{\hat{a}^n, \tilde{s}_n, \hat{c}^n, \hat{h}^n\}_{n\in\mathcal{N}}^{[t+1]}$ by solving problem (6.30) and transmits them to MEC server;

   b) The MEC server updates the global variables $\{a, \tilde{c}, h\}^{[t+1]}$ and transmits them to each SeNB;

   c) The MEC server updates the Lagrange multipliers $\{\sigma^n, \omega^n, \tau^n\}_{n\in\mathcal{N}}^{[t+1]}$ and transmits them to each SeNB; $t=t+1$.

   Until $\| a^{[t+1]} - a^{[t]} \|_2 \leq \vartheta_{dual}$, $\| \tilde{c}^{[t+1]} - \tilde{c}^{[t]} \|_2 \leq \vartheta_{dual}$ and $\| h^{[t+1]} - h^{[t]} \|_2 \leq \vartheta_{dual}$.

3: Output

   Output the optimal solution $\{a, \tilde{s}, \tilde{c}, h\}^*$.

UE, and we employ the primal-dual interior-point method for problem solving in each iteration; thus the complexity would be $\mathcal{O}(k^x)$ with $x > 0$. In global variables updating (6.24)–(6.26), we denote y as the number of elementary steps needed for calculation in (6.24)–(6.26). Then the complexity is given as kNy. In Lagrange multipliers updating (6.27)–(6.29), we use z as the number of elementary steps needed for calculation in (6.27)–(6.29), so the time complexity is calculated as kz. Thus the sum of time complexity in each iteration would be $\mathcal{O}(k^x) + kNy + kz = \mathcal{O}(k^x)$. Assuming P stands for the number of iterations needed for algorithm convergence, the overall time complexity of the distributed algorithm would be $\mathcal{O}(k^x)P$. As will be shown in simulation, the number of iterations before algorithm convergence is not large. So it can be seen that our proposed distributed algorithm can significantly reduce the time complexity compared to the centralized algorithm.

Our overall resource allocation algorithm is summarized in Algorithm 5.

Table 6.2: Simulation parameters

Parameter	Value
Bandwidth	20MHz
Transmission power of UE n, P_n	100 mWatts
Background noise σ^2	-100 dBm
Data size for computation offloading Z_{k_n}	420 KB
Number of CPU cycles of computation task D_{k_n}	1,000 Megacycles
Computation capability of UE n, $f_{k_n}^{(l)}$	0.7 GHz [47]
Computation capability of the MEC server F	100 GHz [47]

6.5 Simulation results and discussion

In this section, simulation results of the proposed decentralized scheme are presented in comparison with the centralized scheme and several baseline schemes. The simulation is run on a MATLAB-based simulator. Unless otherwise mentioned, most of the simulations employ the following scenario. We consider 10 to 50 small cells that are randomly deployed in a 120×120 m^2 area. It is worth noting that most of the results of simulation studies in this section are based on an average over a number of Monte Carlo simulations for various system parameters. There are 4 to 10 UE connected to one SeNB, as mentioned in Section 6.2. The transmission power of a single UE, P_n, is set to 100 mW. The channel gain models presented in 3GPP standardization are adopted here. The total size of the Internet content is 1000 files, and the storage capability of the MEC server is 1000 files. The main simulation parameters employed in the simulations, unless mentioned otherwise, are summarized in Table 6.2.

We first present the convergence of the proposed ADMM-based algorithm with different values of parameter ρ. As shown in Figure 6.2, the utilities of the ADMM-based algorithm increase dramatically in the first 15 iterations and then enter a stable status within the first 40 iterations. So it is proper to say that the decentralized algorithm can converge quickly. All three iterative progresses converge to the same utility value eventually. The progress with a ρ value $\rho = 1.2$ converges fastest, while $\rho = 0.4$ slowest, but the difference is not significant. As can be seen in Figure 6.2, the gap between the ADMM-based algorithm and the centralized algorithm is narrow.

Next the percentages of offloading UE in all UE with an increasing total number of small cells are shown in Figure 6.3. Here the number of UE connected to each SeNB is set as 6. The offloading UE percentages of the ADMM-based algorithm and the centralized algorithm are compared in Figure 6.3. The percentage of offloading UE remains 100 with a small total number of small cells, but as the total number of small cells keeps increasing, the percentage of offloading UE begins to decline. This is because when the total number of small cells is small enough, all the UE are allowed to offload

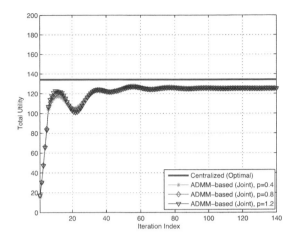

Figure 6.2: Convergence progress of the ADMM-based algorithm with different values of ρ.

their computation tasks in order to maximize network revenue; on the other hand, when the total number of small cells becomes large enough, there will be more UE that tend to offload their computation tasks to the MEC server, and this would cause severe interference with each other, so the algorithm automatically rejects some of the offloading requests generated by UE.

Figure 6.4 and Figure 6.5 show the spectrum and computation resource distribution among all the UE in the system, respectively. Here we only set 4 small cells in system, and there are 4 UE associating to each SeNB. In

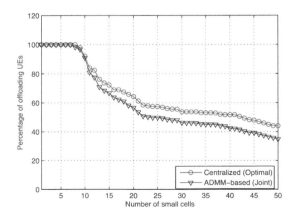

Figure 6.3: Percentage of offloading UE versus number of small cells.

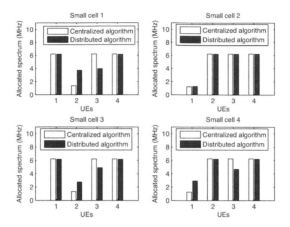

Figure 6.4: Spectrum allocation among UE.

this case all 16 UEs are allowed to offload their computation tasks to the MEC server. As we can see, due to different channel conditions and sizes of computation tasks of different UE, the resource allocation among UE is not uniform, in order to reach the optimal utility value. In Figure 6.4 and Figure 6.5, the resource allocation decisions of our ADMM-based algorithm are compared with that of the centralized algorithm. As shown, except for slight discrepancies on a few UE, they approximately coincide with each other.

Figure 6.6 shows the total alleviated backhaul usage with respect to an increasing number of small cells. Note that total alleviated backhaul means the accumulated alleviation of backhaul usage upon all the UE in system. As

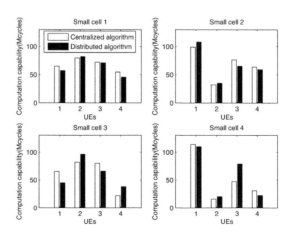

Figure 6.5: Computation resource allocation among UE.

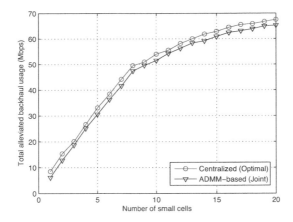

Figure 6.6: Total alleviated backhaul usage versus number of small cells.

shown in Figure 6.6, the alleviated backhaul of our proposed ADMM-based algorithm is very close to that of the centralized algorithm. The alleviated backhaul of both algorithms keeps increasing with an increasing number of small cells (UE).

Figure 6.7 shows the revenue of the MSO (utility value) with respect to an increasing computational capability, where the total number of ScNBs is 20. The revenue of the ADMM-based algorithm is shown by the red line, in comparison with the revenue of the centralized algorithm (blue line) and other benchmark solutions (ADMM solution without caching, ADMM solution with spectrum uniformly allocated, ADMM solution with computation resource

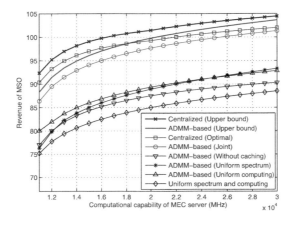

Figure 6.7: MSO revenue versus MEC server computational capability.

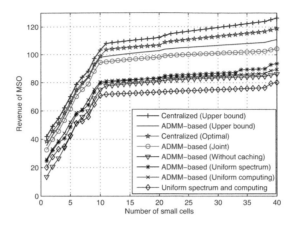

Figure 6.8: MSO revenue versus the number of small cells.

uniformly allocated, spectrum and computation resource uniformly allocated, respectively). In Figures 6.7, 6.8 and 6.9, we also calculate the revenue before the binary variables are recovered and present them, which serve as the upper bound results and are given the legend "Upper bound" to demonstrate that the binary variables recovery operation does not significantly reduce algorithm performance. The centralized algorithm achieves the highest revenue among all the solutions, but it is obvious that the gap between the ADMM-based algorithm and the centralized algorithm is not wide. In contrast, the ADMM with the uniform spectrum allocation solution and the ADMM with the uniform computation resource allocation solution can just achieve much lower revenue. This is because uniform resource allocation usually cannot reach optimal revenue. Then it is no wonder the solution with uniform spectrum and computation resource allocation achieves the lowest MSO revenue. Finally, without the revenue of alleviated backhaul bandwidth, the ADMM solution without caching can only achieve a much lower total revenue compared to the joint ADMM-based solution.

Figure 6.8 shows the revenue of the MSO (utility value) with respect to the increasing number of small cells. It can be seen that with an increasing number of small cells, the revenues of all the solutions increase dramatically at first, because with more and more small cells joining the system, the spectrum can be reused among more and more small cells. Then the MSO could gain more from allocating spectrum to more small cells. Nevertheless, when the number of small cells reaches about 10, the acceleration of the increasing revenue significantly goes down. The main reason is that when there are too many small cells in the system, all those UE will cause severe interference with each other during the computation task offloading process, so the algorithm automatically declined some of the offloading requests generated by UE. Besides, the

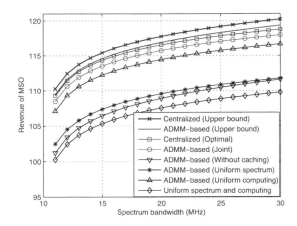

Figure 6.9: MSO revenue versus the available bandwidth.

computation resource of the MEC server cannot be reused at the same time, so if there are too many offloading UE, the amount of computation resource assigned to each UE will decrease, which implies that the MSO will gain less from a single UE. The centralized algorithm and our proposed ADMM-based algorithm achieve relatively high revenue among all the six solutions. Again, because of the uniform allocation of spectrum and computation resources, the other three solutions in which uniform resource allocation strategies are adopted can just achieve much lower revenue. Similarly, due to lack of revenue from alleviated backhaul bandwidth, the revenue of the ADMM solution without caching is also low.

In Figure 6.9, the revenue of ADMM is compared with those of the centralized algorithm and the other four baseline solutions. In this figure, the number of small cells is set as 20. It can be seen from Figure 6.9 that the revenue of our proposed ADMM algorithm is close to the revenue achieved by the centralized algorithm under various spectrum bandwidth conditions. The solution of ADMM with uniform computation resource allocation but optimal spectrum allocation can achieve relatively higher revenue compared with the other two uniform solutions (ADMM with uniform spectrum allocation and uniform spectrum and computation resource allocation). This is mainly because that spectrum allocation usually plays a more important role in earning the interest of the MSO, and this in turn is due to the fact that unlike computation resources of the MEC server, the spectrum resource could be reused simultaneously among UE in different small cells.

Next we discuss the average UE time consumption in the system for executing computation tasks, and the execution time expression proposed in Section 6.2.3.2 is employed here to present the results. Figure 6.10 shows the average UE time consumption of the ADMM-based algorithm compared with

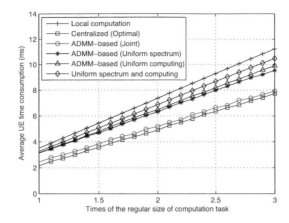

Figure 6.10: Average UE time consumption versus the size of the computation task.

that of the centralized algorithm, local computation solution and the other three baseline solutions. The number of small cells here is 20. The y axis is the average time consumption of all the UE in system and the x axis is the size of computation tasks W_{k_n} (including Z_{k_n} and D_{k_n}), expressed in time of the regular size of computation task. Without losing generality, different UE are considered to have different regular sizes of computation tasks, but the regular sizes are all around the values shown in Table 6.2. The largest time consumption is achieved by the local computation solution, in which all the UE in the system execute their computation tasks on local devices. Because of the shortage of computational resources, the local computation solution consumes much time, especially when the size of the computation task is large. The benchmark solution in which both MEC computation and spectrum resources are uniformly allocated achieves less time consumption compared with the local computation solution. This is due to the fact that the MEC server is more powerful than the UE devices. Because resource allocation is not optimized, the advantage over the local computation solution is not significant. ADMM with uniform computation resource allocation and ADMM with uniform spectrum resource allocation consume less time than the two solutions discussed above. The centralized algorithm consumes the least time among all the solutions. But as we can see, the gap between the centralized algorithm and our proposed ADMM-based algorithm is narrow.

6.6 Conclusions and future work

In this chapter, we presented an ADMM-based decentralized algorithm for computation offloading, resource allocation and Internet content caching optimization in heterogeneous wireless cellular networks with mobile edge computing. We formulated the computation offloading decision, spectrum resource allocation, MEC computation resource allocation, and content caching issues as an optimization problem. Then, in order to tackle this problem in an efficient way, we presented an ADMM-based distributed solution, followed by a discussion about the feasibility and complexity of the algorithm. Finally, the performance evaluation of the proposed scheme was presented in comparison with the centralized solution and several baseline solutions. Simulation results demonstrated that the proposed scheme can achieve better performance than other baseline solutions under various system parameters. Future work is in progress to consider wireless network virtualization in the proposed framework.

References

[1] S. Abolfazli, Z. Sanaei, E. Ahmed, A. Gani, and R. Buyya, "Cloud based augmentation for mobile devices: Motivation, taxonomies, and open challenges," *IEEE Commun. Surveys Tuts.*, vol. 16, no. 1, pp. 337–368, 2014.

[2] R. Xie, F. R. Yu, H. Ji, and Y. Li, "Energy–efficient resource allocation for heterogeneous cognitive radio networks with femtocells," *IEEE Trans. Wireless Commun.*, vol. 11, no. 11, pp. 3910–3920, Nov. 2012.

[3] S. Bu and F. R. Yu, "Green cognitive mobile networks with small cells for multimedia communications in the smart grid environment," *IEEE Trans. Veh. Technol.*, vol. 63, no. 5, pp. 2115–2126, Jun. 2014.

[4] G. Huang and J. Li, "Interference mitigation for femtocell networks via adaptive frequency reuse," *IEEE Trans. Veh. Technol.*, vol. 65, no. 4, pp. 2413–2423, Apr. 2016.

[5] A. R. Elsherif, W.–P. Chen, A. Ito, and Z. Ding, "Adaptive resource allocation for interference management in small cell networks," *IEEE Trans. Commun.*, vol. 63, no. 6, pp. 2107–2125, Jun. 2015.

[6] Y. Cai, F. R. Yu, and S. Bu, "Cloud computing meets mobile wireless communications in next generation cellular networks," *IEEE Netw.*, vol. 28, no. 6, pp. 54–59, Nov. 2014.

[7] Z. Yin, F. R. Yu, S. Bu, and Z. Han, "Joint cloud and wireless networks operations in mobile cloud computing environments with telecom operator cloud," *IEEE Trans. Wireless Commun.*, vol. 14, no. 7, pp. 4020–4033, Jul. 2015.

[8] M. Zhanikeev, "A cloud visitation platform to facilitate cloud federation and fog computing," *Computer*, vol. 48, no. 5, pp. 80–83, May 2015.

[9] S. Sarkar, S. Chatterjee, and S. Misra, "Assessment of the suitability of fog computing in the context of Internet of Things," *IEEE Trans. Cloud Comput.*, vol. 6, no. 1, pp. 46–59, Oct. 2018.

[10] R. Gargees et al., "Incident–supporting visual cloud computing utilizing software–defined networking," *IEEE Trans. Circuits Syst. Video Technol.*, vol. 27, no. 1, pp. 182–197, Jan. 2017.

[11] M. Patel et al., "Mobile–edge computing: Introductory technical white paper," *ETSI, Sophia Antipolis, France, White Paper V1 18–09–14*, Sep. 2014.

[12] J.-Q. Li, F. R. Yu, G. Deng, C. Luo, Z. Ming, and Q. Yan, "Industrial Internet: A survey on the enabling technologies, applications, and challenges," *IEEE Commun. Surveys Tuts.*, vol. 19, no. 3, pp. 1504–1526, Apr. 2017.

[13] O. Kinen, "Streaming at the edge: Local service concepts utilizing mobile edge computing," in *Proc. 9th IEEE Int. Conf. Next Generat. Mobile Appl. Services Technol.*, Cambridge, U.K., Sep. 2015, pp. 1–6.

[14] Y.-D. Lin, E. T.-H. Chu, Y.-C. Lai, and T.-J. Huang, "Time–and–energy aware computation offloading in handheld devices to coprocessors and clouds," *IEEE Syst. J.*, vol. 9, no. 2, pp. 393–405, Jun. 2015.

[15] O. Munoz, A. Pascual–Iserte, and J. Vidal, "Optimization of radio and computational resources for energy efficiency in latency–constrained application offloading," *IEEE Trans. Veh. Technol.*, vol. 64, no. 10, pp. 4738–4755, Oct. 2015.

[16] X. Chen, "Decentralized computation offloading game for mobile cloud computing," *IEEE Trans. Parallel Distrib. Syst.*, vol. 26, no. 4, pp. 974–983, Apr. 2015.

[17] Y. He, F. R. Yu, N. Zhao, H. Yin, H. Yao, and R. C. Qiu, "Big data analytics in mobile cellular networks," *IEEE Access*, vol. 4, pp. 1985–1996, Mar. 2016.

[18] S. Deng, L. Huang, J. Taheri, and A. Y. Zomaya, "Computation offloading for service workflow in mobile cloud computing," *IEEE Trans. Parallel Distrib. Syst.*, vol. 26, no. 12, pp. 3317–3329, Dec. 2015.

[19] C. Fang, F. R. Yu, T. Huang, J. Liu, and Y. Liu, "A survey of green information–centric networking: Research issues and challenges," *IEEE Commun. Surveys Tuts.*, vol. 17, no. 3, pp. 1455–1472, 2015.

[20] K. Wang, H. Li, F. R. Yu, and W. Wei, "Virtual resource allocation in software–defined information–centric cellular networks with device–todevice communications and imperfect CSI," *IEEE Trans. Veh. Technol.*, vol. 65, no. 12, pp. 10011–10021, Dec. 2016.

[21] C. Liang, F. Yu, and X. Zhang, "Information–centric network function virtualization over 5g mobile wireless networks," *IEEE Netw.*, vol. 29, no. 3, pp. 68–74, Jun. 2015.

[22] Z. Li, Q. Wu, K. Salamatian, and G. Xie, "Video delivery performance of a large–scale vod system and the implications on content delivery," *IEEE Trans. Multimedia*, vol. 17, no. 6, pp. 880–892, Jun. 2015.

[23] W. Chai, D. He, I. Psaras, and G. Pavlou, "Cache less for more in information–centric networks (extended version)," *Comput. Commun.*, vol. 36, no. 7, pp. 758–770, Apr. 2013.

[24] Y. Wang, Z. Li, G. Tyson, S. Uhlig, and G. Xie, "Optimal cache allocation for content–centric networking," in *Proc. IEEE ICNP*, Oct. 2013, pp. 1–10.

[25] K. Cho, M. Lee, K. Park, T. T. Kwon, Y. Choi, and S. Pack, "WAVE: Popularity–based and collaborative in–network caching for content oriented networks," in *Proc. IEEE INFOCOMM WKSHPS*, Mar. 2012, pp. 316–321.

[26] Z. Chen and D. Wu, "Rate–distortion optimized cross–layer rate control in wireless video communication," *IEEE Trans. Circuits Syst. Video Technol.*, vol. 22, no. 3, pp. 352–365, Mar. 2012.

[27] A. H. Jafari, D. Lpez–Prez, H. Song, H. Claussen, L. Ho, and J. Zhang, "Small cell backhaul: Challenges and prospective solutions," *EURASIP J. Wireless Commun. Netw.*, vol. 2015, no. 1, p. 206, Dec. 2015.

[28] L. Yang, J. Cao, Y. Yuan, T. Li, A. Han, and A. Chan, "A framework for partitioning and execution of data stream applications in mobile cloud computing," *ACM SIGMETRICS Perform. Eval. Rev.*, vol. 40, no. 4, pp. 23–32, Mar. 2013.

[29] G. Iosifidis, L. Gao, and L. Tassiulas, "An iterative double auction mechanism for mobile data offloading," *IEEE/ACM Trans. Netw.*, vol. 23, no. 5, pp. 1634–1647, Oct. 2015.

[30] L. Ma, F. Yu, V. C. M. Leung, and T. Randhawa, "A new method to support UMTS/WLAN vertical handover using SCTP," *IEEE Wireless Commun.*, vol. 11, no. 4, pp. 44–51, Aug. 2004.

[31] L. Ma, F. R. Yu, and V. C. M. Leung, "Performance improvements of mobile SCTP in integrated heterogeneous wireless networks," *IEEE Trans. Wireless Commun.*, vol. 6, no. 10, pp. 3567–3577, Oct. 2007.

[32] F. Yu and V. Krishnamurthy, "Optimal joint session admission control in integrated WLAN and CDMA cellular networks with vertical handoff," *IEEE Trans. Mobile Comput.*, vol. 6, no. 1, pp. 126–139, Jan. 2007.

[33] S. Bu, F. R. Yu, Y. Cai, and X. P. Liu, "When the smart grid meets energy–efficient communications: Green wireless cellular networks powered by the smart grid," *IEEE Trans. Wireless Commun.*, vol. 11, no. 8, pp. 3014–3024, Aug. 2012.

[34] B.-G. Chun, S. Ihm, P. Maniatis, M. Naik, and A. Patti, "CloneCloud: Elastic execution between mobile device and cloud," in *Proc. EuroSys*, Salzburg, Austria, Apr. 2011, pp. 301–314.

[35] X. Wang, M. Chen, T. Taleb, A. Ksentini, and V. C. M. Leung, "Cache in the air: Exploiting content caching and delivery techniques for 5G systems," *IEEE Commun. Mag.*, vol. 52, no. 2, pp. 131–139, Feb. 2014.

[36] M. Fiore, F. Mininni, C. Casetti, and C.-F. Chiasserini, "To cache or not to cache?" in *Proc. IEEE INFOCOM*, Apr. 2009, pp. 235–243.

[37] C. Fricker, P. Robert, J. Roberts, and N. Sbihi, "Impact of traffic mix on caching performance in a content–centric network," in *Proc. IEEE Conf. Comput. Commun. Workshops (INFOCOM WKSHPS)*, Mar. 2012, pp. 310–315.

[38] N. Golrezaei, K. Shanmugam, A. G. Dimakis, A. F. Molisch, and G. Caire, "Femtocaching: Wireless video content delivery through distributed caching helpers," in *Proc. IEEE INFOCOM*, Mar. 2012, pp. 1107–1115.

[39] D. Bethanabhotla, O. Y. Bursalioglu, H. C. Papadopoulos, and G. Caire, "User association and load balancing for cellular massive MIMO," in *Proc. Inf. Theory Appl. Workshop (ITA)*, Feb. 2014, pp. 1–10.

[40] D. Fooladivanda and C. Rosenberg, "Joint resource allocation and user association for heterogeneous wireless cellular networks," *IEEE Trans. Wireless Commun.*, vol. 12, no. 1, pp. 248–257, Jan. 2013.

[41] S. Gortzen and A. Schmeink, "Optimality of dual methods for discrete multiuser multicarrier resource allocation problems," *IEEE Trans. Wireless Commun.*, vol. 11, no. 10, pp. 3810–3817, Oct. 2012.

[42] S. Boyd and L. Vandenberghe, *Convex Optimization*. Cambridge, U.K.: Cambridge Univ. Press, 2009.

[43] S. Boyd, N. Parikh, E. Chu, B. Peleato, and J. Eckstein, "Distributed optimization and statistical learning via the alternating direction method of multipliers," *Found. Trends Mach. Learn.*, vol. 3, no. 1, pp. 1–122, Jan. 2011.

[44] L. Chen, F. R. Yu, H. Ji, G. Liu, and V. C. M. Leung, "Distributed virtual resource allocation in small–cell networks with full–duplex selfbackhauls and virtualization," *IEEE Trans. Veh. Technol.*, vol. 65, no. 7, pp. 5410–5423, Aug. 2016.

[45] W.-C. Liao, M. Hong, H. Farmanbar, X. Li, Z.-Q. Luo, and H. Zhang, "Min flow rate maximization for software defined radio access networks," *IEEE J. Sel. Areas Commun.*, vol. 32, no. 6, pp. 1282–1294, Jun. 2014.

[46] M. Leinonen, M. Codreanu, and M. Juntti, "Distributed joint resource and routing optimization in wireless sensor networks via alternating direction method of multipliers," *IEEE Trans. Wireless Commun.*, vol. 12, no. 11, pp. 5454–5467, Nov. 2013.

[47] X. Chen, L. Jiao, W. Li, and X. Fu, "Efficient multi–user computation offloading for mobile–edge cloud computing," *IEEE/ACM Trans. Netw.*, vol. 24, no. 5, pp. 2795–2808, Oct. 2015.

Chapter 7

Software-Defined Networking, Caching and Computing

In this chapter, we propose a novel framework that jointly considers networking, caching and computing techniques in order to improve end-to-end system performance. This integrated framework can enable dynamic orchestration of networking, caching and computing resources to meet the requirements of different applications. We define and develop the key components of this framework: data, control and management planes. The data plane consists of the devices that are responsible for networking, caching and computing operations. The control plane has a logically centralized controller to guide these operations. The management plane enables not only traditional applications, such as traffic engineering, but also new applications, such as content distribution and big data analytics. Based on this framework, we consider the bandwidth, caching and computing resource allocation issue and formulate it as a joint caching/computing strategy and server selection problem to minimize the combination cost of network usage and energy consumption. Then we solve it using an exhaustive-search algorithm. Simulation results show that our proposed framework significantly outperforms the traditional network without in-network caching/computing in terms of network usage and energy consumption. In addition, we discuss a number of challenges of implementing the proposed framework of software-defined networking, caching and computing.

7.1 Introduction

Recent advances in information and communications technologies have fueled a plethora of innovations in various areas, including *networking, caching* and *computing*, which can have profound impacts on our society through the development of smart cities, smart transportation, smart homes, etc. In the area of networking, *software-defined networking* (SDN) has attracted great interest from both academia and industry [1]. The basic principle of SDN is to introduce the ability of programming the network via a logically software-defined controller, and separate the control plane from the data plane. SDN allows logical centralization of control, and decisions are made by the centralized controller, which has a global view of the network. Compared to traditional networking paradigms, SDN makes it easier to introduce new abstractions in networking, simplifying network management and facilitating network evolution.

Another new technology called *information-centric networking* (ICN) has been extensively studied in recent years [2, 3]. *In-network caching* is used in ICN to reduce duplicate content transmission in networks. ICN-based caching has been recognized as one of the promising techniques for future wireless/wired networks. By promoting content to a first-class citizen in the network, ICN provides native support for scalable and highly efficient content retrieval while enabling an enhanced capability for mobility and security.

At the front of computing, the *cloud computing* paradigm has been widely adopted to enable convenient, on-demand network access to a shared pool of configurable computing resources [4]. Nevertheless, as the distance between the cloud and the edge device is usually large, cloud computing services may not provide guarantees to low latency applications, and transmitting a large amount of data (e.g., in big data analytics) from the device to the cloud may not be feasible or economical. To address these issues, *fog computing* has been proposed to deploy computing resources closer to end devices [5, 6]. A similar technique, called *mobile edge computing*, is being standardized to allocate computing resources in wireless access networks [7].

Although some excellent work has been done on networking, caching and computing, these three important areas have traditionally been addressed separately in the literature. However, as shown in the following, it is necessary to jointly consider these three advanced techniques to provide better service in future networks. Therefore, in this chapter, we propose to jointly consider networking, caching and computing techniques in order to improve end-to-end system performance. The motivations behind our work are based on the following observations.

- From the perspective of applications (e.g., video streaming), network, cache and compute are underlying resources enabling these applications. How to manage, control and optimize these resources can have significant impacts on the performance of applications.

- In the existing works of SDN, ICN, and cloud/fog computing, these resources are managed, controlled and optimized separately, which could result in suboptimal performance.

■ Therefore, an integrated framework for networking, caching and computing has the potential to significantly improve the end-to-end performance of applications.

The main contributions of this chapter are as follows.

■ We propose a novel framework called SD-NCC (Software-Defined Networking, Caching and Computing) that integrates networking, caching and computing in a systematic way to naturally support data retrieval and computing services. Based on the programmable control principle originated in SDN, we incorporate the ideas of information centricity originated in ICN. This integrated framework can enable dynamic orchestration of networking, caching and computing resources to meet the requirements of different applications.

■ We define and develop the key components of this framework: data, control and management planes. The data plane consists of the devices that are responsible for networking, caching and computing operations. The control plane has a logically centralized controller to guide these operations. The management plane enables not only traditional applications, such as traffic engineering, but also new applications, such as content distribution and big data analytics.

■ In essence, it is the unique dynamics tied with networking, caching and computing that present interesting challenges beyond the current works. We believe that these initial steps we have taken here open a new avenue for an integrated framework for software-defined networking, caching and computing, and shed light on effective and efficient designs therein.

■ We present a comprehensive system model for the SD-NCC framework. Then we formulate a joint caching/computing strategy and server selection problem as a mixed-integer nonlinear programming (MINLP) problem to minimize the combination cost of network usage and energy consumption in the framework. Specifically, we formulate the caching/computing capacity allocation problem and derive the optimal deployment numbers of service copies. In addition, we develop two algorithms to find the optimal caching/computing strategy and server selection strategy.

■ We conduct simulations to evaluate system performance and deployment cost by comparing our proposed framework with the traditional network without in-network caching/computing. Our proposed SD-NCC framework significantly reduces the traffic traversing the network and has a performance advantage on energy consumption cost. We also show the impact of service popularity on the optimal deployment number of service copies.

7.2 Recent advances in networking, caching and computing

Networking, caching and computing have traditionally been addressed separately. In this section, we briefly present recent advances in each area. The challenges of these areas will be discussed as well.

7.2.1 Software-defined networking

SDN [1] decouples the control plane (decision functions) and the data plane (forwarding functions) of a network and brings high programmability to network devices. This feature enables easier realization for complex network services without the need to replace hardware components in the network. It also makes the network easier to configure, manage, troubleshoot and debug. The controller installs control demands to the devices in the data plane by means of a well-defined application programming interface (API) between them. The most notable example of such an API is OpenFlow [8]. An SDN-enabled device is instructed by the controller via flow tables, each of which defines a subset of the traffic and the corresponding action. Thus an SDN-enabled device can behave like a router, firewall or other middle-box.

SDN has been considered one of the most promising technologies on realization of programmable network control in both wire and wireless networks [9, 10, 11], and has been deployed well in existing IP networks, such as Internet service providers and data center networks. Nevertheless, there are some challenges of SDN as follows.

- Currently, SDN mainly focuses on controlling the forwarding function of switching devices. Consequently, non-forwarding capabilities (e.g., caching and computing) of the devices are largely ignored in SDN. However, these non-forwarding capabilities can have significant impacts on system performance, and they could be controlled by the control plane for various programmable policies to improve performance.

- In SDN, since the flow is defined as a combination of several matched fields in the header of packets (such as MAC and IP addresses, and protocol ports), it is difficult for per-flow based forwarding behaviors to naturally support different forwarding policies for different content objects (i.e., a finer granularity of the programmable policy).

- The network function provided by the control plane in SDN is coupled with the node location, which is no longer fixed with the increasingly use of virtualization and the prevalence of mobility.

7.2.2   Information centric networking

ICN [2] is a clean-slate approach to evolve the Internet infrastructure away from host-centric end-to-end communication to receiver-driven content retrieval based on "named information" (or content or data). Content centric networking (CCN) [12] is one specific proposal of ICN. It keeps the successful TCP/IP principles "thin waist" at the network layer, but places data's names instead of their locations at this layer. In addition, it integrates fundamental architectural primitives: building security into data itself, inherent self-regulating network traffic, and adaptive routing and forwarding capabilities.

Communication in CCN is driven by data consumers through the exchange of two types of packets: Interest and Data. In CCN, data consumers issue Interest packets to request information objects, which return in the form of Data packets, with both types of packets carrying a name that identifies the requested information object. A data packet in response to an Interest packet traverses exactly the same links in the reverse order. CCN nodes have three basic components: a Pending Interest Table (PIT), a Forwarding Information Base (FIB) and a Content Store (CS). The PIT stores information about Interests that a router has forwarded but not satisfied yet. Each PIT entry records the name carried in the Interest, together with its incoming and outgoing interface(s). FIB is populated by a name-prefix based routing protocol and can have multiple output interfaces for each prefix. CS is a temporary cache of Data packets that the router has received.

Despite the promising features of ICN, this clean-slate approach is not easy for network operators to deploy, since it may require replacement or update of existing equipment. In addition, there are some key challenges that are fundamental to globally scale the CCN architecture, such as name-space organization in naming design, key and trust management and a scalable forwarding plane.

7.2.3   Cloud and fog computing

Recent advances in cloud computing provide a range of services to enterprises and end-users, freeing them from the specification of many details, such as computation and storage resources [4, 13]. The basic concept behind cloud computing is to offload the users' data and computations to remote resource providers, such as third-party data centers. Although widely adopted, this paradigm, which utilizes the storage/computing capabilities in a remote provider's servers over the Internet backbone, can hardly satisfy some applications' requirements: such as mobility support, location awareness and low latency. Besides, it brings a large amount of traffic between the cloud and the end-user.

Fog computing, a paradigm that extends cloud computing services close to the user (edge devices and even the devices themselves common in Internet of Things scenarios), has been proposed in the last few years, and claims the solutions to these limitations [6]. It's a distributed infrastructure using a multitude of end-users or near-user edge devices to carry out a collaborative and distributed service. It is distinguished from the cloud by its proximity to end-users, its support for mobility and the dense geographical distribution of local resource pooling. Nevertheless, there are also some issues that hinder its wide deployment, such as the limitation of battery, storage, computing and information awareness in devices, the challenges of management and security, and the high level of volatility in the infrastructure created in temporary fog computing.

7.2.4 An integrated framework for software-defined networking, caching and computing

In this chapter, we present a novel framework that integrates networking, caching and computing in a systematic way to naturally support data retrieval and computing services. Based on the programmable control principle originated in SDN, we incorporate the ideas of information centricity originated in ICN. Moreover, we take into account the computing capabilities of network elements in this framework. In the following, we highlight three important features of SD-NCC.

7.2.4.1 Software-defined and information-centric control

Similar to SDN, the control plane is a software orchestration platform facilitating software applications to interact with the underlying data plane devices. It transfers the policy defined in the management plane to control logic and then installs the control commands on the underlying data plane devices. By naming data instead of naming an address, the forwarding and processing operations of network devices are based on names in packets. Thus, compared with SDN, SD-NCC takes a finer-grained flow as an operation unit, enabling an information-centric paradigm.

7.2.4.2 Service-oriented request/reply paradigm

Similar to ICN, communication in SD-NCC is driven by service consumers through the exchange of two types of packets: Interest and Data. Both types of packets carry a hierarchical and unique service name. The service names of SD-NCC packets cover broadly according to different application purposes. Here, we focus on the two kinds of services: content service and computing service, which are the major requirements from users. To request a service, a consumer puts the name of the desired service into an Interest packet and sends it to the network. Network devices use this name to forward the Interest to the service producer(s). Once the Interest reaches a node that can provide service, the node will return a Data packet that contains both the name and the data. One Data satisfies one Interest across each hop and Data takes the exact same path as the Interest that requested it, but in the reverse direction.

7.2.4.3 In-network caching and computing

In-network caching and computing become inherent fundamental capabilities in SD-NCC. Additionally, each SD-NCC Data packet is meaningful independent of where it comes from or who requests it, since it carries a name and a signature. Thus, Data packets that have content can be cached in any devices to serve subsequent content requests. Moreover, a computation service requirement will be satisfied in the intermediate device that has in-network computing capability. These ubiquitous and fine-grained capabilities provide significant and important advantages of SD-NCC.

7.3 Architecture of the integrated framework SD-NCC

As Figure 7.1 shows, the architecture of the framework can be divided into three planes of functionality: data, control and management planes. The data plane contains the devices that are responsible for performing a set of elementary operations, including networking, caching and computing, guided by the controller in the control plane. The control plane has a logically centralized controller to control these operations by populating the flow tables of the data plane elements. The management plane includes software services, used to remotely monitor and configure the control functionality, such as routing strategy, load balancing and energy efficiency. These planes are described in detail as follows.

7.3.1 The data plane

The data plane in SD-NCC is composed of a set of networking equipment (such as switches, routers and firewalls), which are interconnected by wired cables or wireless radio channels. It provides original information to the control plane and executes the control command from the controller.

Despite the absence of complicated deep packet inspection (DPI) equipment to identify the value content from the enormous flow, the data plane in our architecture is information-aware. By naming data instead of naming an address, the forwarding and processing operations of network devices are based on names in packets. From these named packets, the data plane devices can naturally obtain a large number of information resources.

Another main difference resides in the fact that those physical devices are now embedded with control to take autonomous decisions instead of just simply forwarding elements. In other words, the network intelligence is not

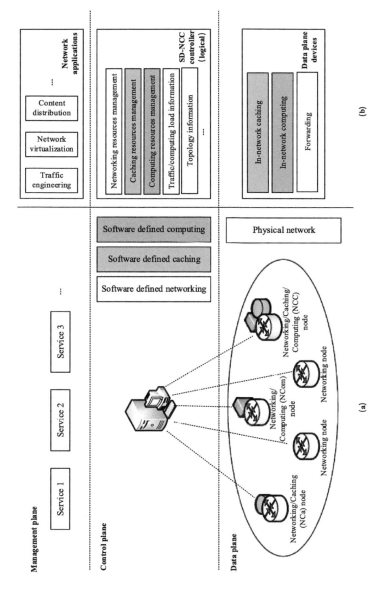

Figure 7.1: Software-defined networking, caching and computing in (a) planes and (b) system design architecture.

totally removed from the data plane element to a logically centralized control plane. However, it's important to note that the local strategies in the data plane devices are also preset by the controller.

The data plane devices use the service name to perform a set of elementary operations according to the flow tables installed by the control plane. As Figure 7.2 shows, through the southbound API, a pipeline of flow tables in the data plane devices guides forwarding and processing behavior. When there are no matched rules for an arriving packet, the local processing pipeline begins to take effect. Note that this local processing strategy represents the intelligence of the data plane. Figure 7.3 and Figure 7.4 illustrate how to handle the Data and Interest packets in detail. Interest packets and Data packets go through different kinds of processes. The local pipeline of Interest processing includes three parts: local caching pipeline, local computing pipeline and local forwarding pipeline. According to the local strategy, there will be a service names list of requests which are chosen to be served on a local node (caching and computing services). As Figure 7.3 shows, upon receiving an Interest packet, the device looks up the local services list based on the service name of the incoming Interest. If the service name of this Interest packet is on the list, it will be sent to the local caching or computing pipeline according to the service type; otherwise, it will be sent to the local forwarding pipeline. In the local caching pipeline, the request content is sent back as a Data packet. In the local computing pipeline, the computation task is done and the result is sent back as a Data packet. The local forwarding pipeline has components of Pending Interest Table (PIT) and Forwarding Information Base (FIB), similar to the Interest forwarding mechanism of CCN. First, the device looks up the name in its PIT. If a matching entry exists, it will simply record the incoming interface of this Interest in the PIT entry. In the absence of a matching PIT entry, the device will create a new PIT entry and forward the Interest according to the FIB and local forwarding strategy. If there is no matching FIB entry, the device will encapsulate and forward this Interest to the controller. Figure 7.4 shows the local pipeline of Data processing based on the PIT entry, which is similar to the Data forwarding mechanism in CCN. Upon receiving a Data packet, if the device finds the matching PIT entry, the Data will be forwarded to all downstream interfaces listed in this PIT entry. Then the device will remove this PIT entry and update the CS (whether or not to cache Data depends on the caching strategy). As such, Interest and Data can be successfully forwarded.

7.3.2   The control plane

The control plane is a software orchestration platform that has two essential functions. First, it collects information of network status about the data plane layer offering a global network view to network applications. Second, it transfers the network policy defined in the management plane to control logic and then installs the control commands on the underlying data plane devices.

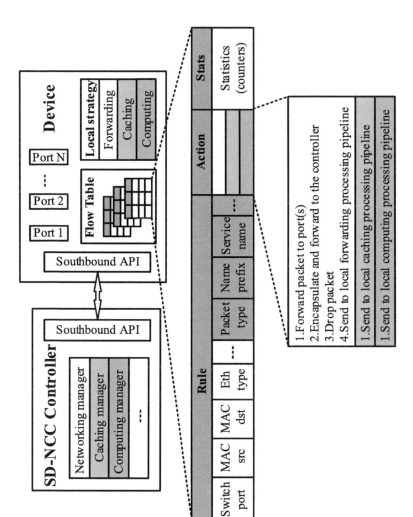

Figure 7.2: The SD-NCC controller and device.

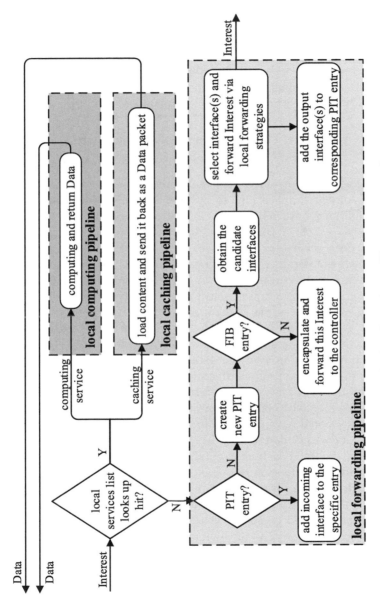

Figure 7.3: The local pipeline of Interest processing.

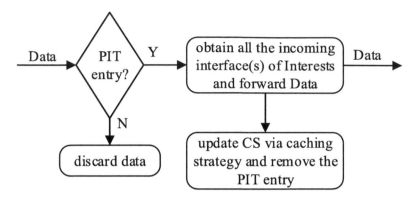

Figure 7.4: The local pipeline of Data processing.

Based on a logically centralized, abstract network view, the control plane facilitates the software applications to interact with the underlying data plane devices, simplifying both policy enforcement and network (re-)configuration and evolution.

Since the data plane layer is no longer just a forwarding layer, the added capabilities (such as information awareness, in-network caching and in-network computing) enable the control plane to hold the more comprehensive underlying network status including physical resources and information resources. For information resources, the control plane collects them from the network devices that are naturally content aware. In contrast to the traditional traffic flow load, the information obtained in SD-NCC is finer-grained. The controller adds a new function to analyze the data information and obtains the behaviors and requirements of users and content providers. For physical resources, besides maintaining the networking resources management as a traditional SDN controller does, the controller in our new framework also manages the two other dimensional resources: caching and computing.

With in-network caching in network devices, the controller collects the information of distributed caching capabilities. Then according to the information analysis from information resources, it can bring popular content close to users and guide the Interest packets, directly forwarding to the closest network device. With in-network computing in network devices, most of users' computing requirements will be served at the intermediate network devices with computing capabilities instead of the remote computing server. On one hand, it decreases the service load on traditional remote servers. On the other hand, the latency of real-time computing services will be significantly decreased. This relies on the logically unified management of computing capabilities in the control plane.

A traditional SDN controller controls forwarding behavior by installing flow-based control commands on forwarding devices. In SD-NCC, we extend the control commands for the functionality of caching and computing, to consider different policies for different services at a finer granularity. As Figure 7.2 shows, the controller installs the control commands by defining a pipeline of flow tables through the southbound API. Each entry of a flow table has three parts: 1) a matching rule; 2) actions to be executed on matching packets; and 3) counters that keep statistics of matching packets. The new matched fields in rules include Packet type, Name prefix and Service name. The Packet type identifies the types of packets. Based on Name prefix and Packet type, forwarding behavior is installed. The forwarding actions are similar to SDN: 1) forward packet to port(s); 2) encapsulate and forward to the controller; 3) drop packet, and 4) send to local processing pipeline. Based on Packet type and Service name, caching and computing behaviors are defined. The corresponding actions are: send to local caching processing pipeline and send to local computation pipeline. When no rule is found for a new arriving packet, a default rule is adopted: send to local processing pipeline.

The newly added management functionalities of caching and computing in the control plane enable SD-NCC to decease redundant traffic and improve utilization of physical resources. Besides, the control plane still maintains the generic functionalities such as network state, network topology information, device discovery, and distribution of network configuration.

7.3.3 The management plane

The management plane running on the top of the control plane defines network policy. It includes a set of software applications that leverage northbound API to implement network operation and control. Based on network information input, there are a wide variety of network applications such as traffic engineering, network virtualization, measurement and monitoring, and mobility. With the additional information of users' requirements and caching/computing capabilities distribution, the management plane also enables some other applications such as big data analytics, content distribution, and distributed computing.

In the big data era, the amount of data from different provisions, such as social networking websites (e.g., Twitter, Facebook, etc.), Internet of Things (IoT) and scientific research, is increasing at an exponential rate. Transmitting a large amount of data from the device to a remote data center for processing may not be feasible or economical. With in-network computing capability, the management plane of SD-NCC can facilitate in-network big data analytics.

With the growth in digital media, e-commerce, social networking, and smartphone applications, content distribution is becoming the dominant use in the current Internet. The mainstream content providers, such as Google, Facebook and Amazon, have their own data centers around the world and build their own content delivery network (CDN) system to accelerate the delivery of web content and rich media to users. In contrast with edge caching, which is the primary technique of a CDN, the in-network caching capability and the programmable control on content routing in SD-NCC enable the management plane to define a suitable content distribution policy to support data transmission in a manipulated and effective way.

With a massive proliferation of computation-intensive applications, computing requirements pose a great challenge on users'/enterprises' devices due to the limitation of power and computing capabilities. By using offloading technology, cloud computing systems help solve this problem. However, they may not provide performance guarantees to low latency applications. With programmable networking and computing capabilities, the management plane in SD-NCC can provide distributed computing services through globally coordinated computing.

7.3.4   The workflow of SD-NCC

We illustrate information transfer and interaction workflows among network nodes (including the controller and data plane devices) for general caching and computing services (Figure 7.5). The main workflows are as follows: 1. The data plane devices, including switches, routers, and routers with additional capabilities, announce their local information to the controller, such as caching, computing capabilities and networking connectivity. The remote server (content provider or computing service provider) also registers with the controller.

2. Based on the information gathered, the controller installs static rules in the flow tables and local strategies of the data plane devices. The static rules include specific forwarding and processing behaviors for each matched packet, while the local strategies for networking, caching and computing are more flexible.

3. An Interest request for content will be forwarded directly to the original destination server when entering the network for the first time. Whether or not to cache the content is decided by the preset rules or local strategies. If the controller decides that the newly cached content needs to register with the controller, subsequent requests will be forwarded to the closer caching node via the newly added rules.

4. An Interest request for computing service will be served in the intermediate device with computing capability, if the controller or local strategy decides so.

5. If an incoming Interest request cannot find the matched rules and FIB entry, the device will encapsulate and forward it to the controller and wait for corresponding rules. Once the rules are created, this Interest packet will be forwarded to the device that can satisfy the request.

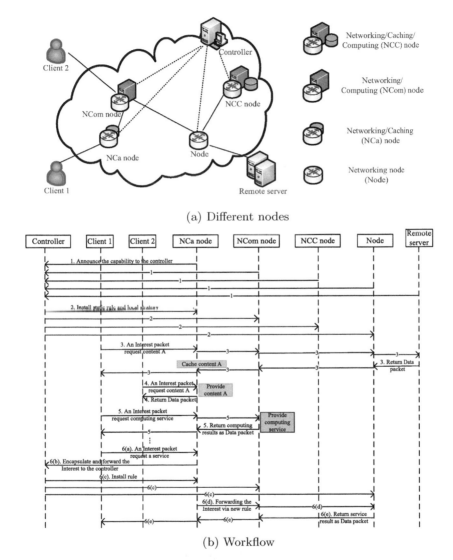

(a) Different nodes

(b) Workflow

Figure 7.5: The workflow of SD-NCC.

7.4 System model

In this section, we present the system model for the SD-NCC framework, including networking, caching, computing and energy models.

7.4.1 *Network model*

Consider a network represented by graph $\mathbf{G} = \{\mathbf{V}, \mathbf{E}\}$, where $\mathbf{V}$ denotes the set of nodes and $\mathbf{E}$ denotes the set of directed physical links. A node can be a router, a user or a server. We introduce the following notation:

- $\mathbf{S} = \{\mathbf{S}_A, \mathbf{S}_B\}$, the set of in-network server nodes; $\mathbf{S}_A \subset \mathbf{V}$, the set of caching nodes; $\mathbf{S}_B \subset \mathbf{V}$, the set of computing nodes.

- o, a single virtual origin node which is always able to serve the service requests.

- $\mathbf{U}$, the set of users who issue service requests, $\mathbf{U} \subset \mathbf{V}$.

- $\mathbf{K} = \{\mathbf{K}_A, \mathbf{K}_B\}$, the set of service elements; $\mathbf{K}_A$, the set of content elements; $\mathbf{K}_B$, the set of computation services.

- $\mathbf{M} : \{m_u^k\}_{(k,u) \in \mathbf{K} \times \mathbf{U}}$, user demand matrix for duration t; m_u^k, the demand of service k from user u, $u \in \mathbf{U}$.

- $(\lambda_{k_1}^A, o_{k_1}^A)$, content k_1's attribute, $k_1 \in \mathbf{K}_A$; $\lambda_{k_1}^A$, the number of content k_1's requests for duration t; $o_{k_1}^A$, the size of content k_1.

- $(\lambda_{k_2}^A, o_{k_2}^B, c_{k_2}^B)$, computation k_2's attribute, $k_2 \in \mathbf{K}_B$; $\lambda_{k_2}^A$, the number of computation k_2's requests for duration t; $o_{k_2}^B$, the amount of computation k_2's data communications, $k_2 \in \mathbf{K}_B$; $c_{k_2}^B$, the amount of computation k_2's workload (which can be measured by the number of clock cycles or execution time).

7.4.2 *Caching/computing model*

Let c_i^A, c_i^B denote the caching capacity and computing capacity of server node i. We assume the finite capacity of in-network server nodes and the infinite capacity of the original server. We use $\mathbf{H} : \{h_i^k\}$ to denote the caching/computing strategy matrix, where $h_i^k = 1$ if server i is able to provide the requested service k and $h_i^k = 0$ if it is not. Specifically, $h_o^k = 1$.

Thus, the caching/computing capacity constraints are as follows:

$$\sum_k h_i^k o_k^A \leq c_i^A , \forall i \in \mathbf{S}_A \tag{7.1}$$

and

$$\sum_k h_i^k c_k^B \leq c_i^B , \forall i \in \mathbf{S}_B \tag{7.2}$$

7.4.3 Server selection model

Let $\mathbf{P} : \{\rho_{i,u}^k\}$ denote the server selection matrix, where $\rho_{i,u}^k \in [0, 1]$ denotes the proportion of the service request k from user u served by server i. Node u can be viewed as an edge router that aggregates the traffic of many end hosts, which may be served by different servers. Let $\mathbf{X} : \{x_{i,u}^k\}$ denote the traffic allocation matrix, where $x_{i,u}^k$ denotes the traffic of service k from user u to server i for duration t. Combining this with the caching/computing strategy, we have

$$x_{i,u}^k = m_u^k \rho_{i,u}^k h_i^k, \forall i \in S \cup \{o\} \tag{7.3}$$

and

$$\sum_i \rho_{i,u}^k h_i^k = 1 \tag{7.4}$$

7.4.4 Routing model

Let $\mathbf{R} : \{r_l^{i,u}\}$ be the routing matrix, where $r_l^{i,u} \in [0, 1]$ denotes the proportion of traffic from user u to server i that traverses link l. $r_l^{i,u}$ equals 1 if the link l is on the path from server i to user u and 0 otherwise.

Then the routing selection strategy is given by

$$\sum_{l:l \in In(v)} r_l^{i,u} - \sum_{l:l \in Out(v)} r_l^{i,u} = I_{v=i}, \ \forall i \in \mathbf{S} \cup \{o\}, v \in \mathbf{V} \backslash \mathbf{U} \tag{7.5}$$

where $I_{v=i}$ is an indicator function which equals 1 if $v = i$ and 0 otherwise, $In(v)$ denotes the set of incoming links to node v, and $Out(v)$ denotes the set of outgoing links from node v.

The capacity of a link l is c_l. Thus, the total traffic traversing link l denoted by x_l is given by

$$x_l = \sum_{k,i,u} x_{i,u}^k r_l^{i,u} \leq c_l, \forall i \in \mathbf{S} \cup \{o\} \tag{7.6}$$

7.4.5 Energy model

Energy is mainly consumed by content caching, data computing and traffic transmission in the SD-NCC framework. We discuss these three energy models as follows.

7.4.5.1 Caching energy

For E_{ca}^A, we use an energy-proportional model, similar to [14]. If n_{k_1} copies of the content k_1 are cached for duration t, the energy consumed by caching $n_k o_k^A$ bits is given by

$$E_{ca,k_1}^A = n_{k_1} o_{k_1}^A p_{ca}^A t \tag{7.7}$$

where p_{ca}^A is the power efficiency of caching. The value of p_{ca}^A strongly depends on the caching hardward technology.

7.4.5.2 Computing energy

We assume that computation applications are executed within the virtual machines (VMs) which are deployed on the computing nodes, and the amount of incoming computation k_2's workloads from all users is $\lambda_{k_2}^B c_{k_2}^B$ for duration t. The energy consumption of the computing nodes consists of two parts, including a dynamic energy consumption part E_{active}^B when the VMs are active (processing computing service requests) and a static energy consumption part E_{static}^B when the VMs are idle.

Thus, the dynamic energy consumption for processing the workloads is

$$E_{active}^{B,k_2} = \lambda_{k_2}^B c_{k_2}^B p_{active}^B \tag{7.8}$$

and the static energy consumption for m_{k_2} copies of computation k_2's dedicated VM is

$$E_{static}^{B,k_2} = m_{k_2} p_{static}^B t \tag{7.9}$$

where p_{active}^B, p_{static}^B are the average power efficiency of VMs in active and static states, respectively.

Note that the dynamic energy consumption is independent of the VMs' copies.

7.4.5.3 Transmission energy

The transmission energy E_{tr} consumption mainly consists of energy consumption at routers and energy consumption along the links.

The transmission energy of content E_{tr}^A and computation E_{tr}^B for duration t is given by

$$F_{tr}^{A,k_1} = \lambda_{k_1}^A o_{k_1}^A \left[p_{tr,link} \cdot d_{k_1}^A + p_{tr,node} \cdot \left(d_{k_1}^A + 1 \right) \right] \tag{7.10}$$

$$E_{tr}^{B,k_2} = \lambda_{k_2}^B o_{k_2}^B \left[p_{tr,link} \cdot d_{k_2}^B + p_{tr,node} \cdot \left(d_{k_2}^B + 1 \right) \right] \tag{7.11}$$

where $p_{tr,link}$ and $p_{tr,node}$ are the energy efficiency parameters of the link and node, respectively. $d_{k_1}^A$ and $d_{k_2}^B$ represent the average hop distance to the content server nodes for content k_1's request and the computation server nodes for computation k_2's request, respectively.

7.5 Caching/computing/bandwidth resource allocation

In this subsection, we consider the joint caching, computing and bandwidth resource allocation problem and formulate it as a joint caching/computing strategy and server selection problem (CCS-SS) which we show is a mixed integer nonlinear programming (MINLP) problem. By relaxing the whole caching, computing, link constraints, we focus on the caching/computing capacity allocation for each service element and formulate it as a nonlinear programming

(NLP) problem. We derive the optimal caching/computing capacity allocation (i.e., the copies number for each service element) and then propose an
exhaustive-search algorithm to find the optimal caching/computing strategy
and server selection.

7.5.1 Problem formulation

7.5.1.1 Objective function

Based on the known topology, traffic matrix, routing matrix and the energy
parameters of network equipment, we introduce the following quantities:

- $d_{i,u}$, the hop distance between server node i and user u;

- $D_{i,u}$, the end-to-end latency between server node i and user u;

- $a_{i,u} = p_{tr,link}d_{i,u} + p_{tr,node}(d_{i,u} + 1)$, the energy for transporting per
  bit from server node i to user u;

- $f_{e,tr} = \sum_{k \in \mathbf{K}} \sum_{i \in \mathbf{S} \cup \{o\}} \sum_{u \in \mathbf{U}} a_{i,u} m_u^k \rho_{i,u}^k h_i^k$, the transmission energy for duration t;

- $f_{e,ca} = \sum_{k_1 \in \mathbf{K}_A} \sum_{i \in \mathbf{S}_A} h_i^{k_1} o_{k_1}^A p_{ca}^A t$, the caching energy for duration t;

- $f_{e,com} = \sum_{k_2 \subset \mathbf{K}_B} \sum_{i \subset \mathbf{S}_B} h_i^{k_2} p_{static}^B t + \sum_{k_2 \in \mathbf{K}_B} \lambda_{k_2}^B c_{k_2}^B p_{active}^B$, the energy consumption on computing nodes for duration t;

- $f_{tr} = \sum_{k \in \mathbf{K}} \sum_{i \in \mathbf{S} \cup \{o\}} \sum_{u \in \mathbf{U}} D_{i,u} m_u^k \rho_{i,u}^k h_i^k$, the network traffic for duration t.

We select a combining objective which balances the energy costs and the
network usage costs for duration t. The cost function is given by

$$f = f_{e,ca}(h_i^k) + f_{e,com}(h_i^k) + f_{e,tr}(\rho_{i,u}^k h_i^k) \\ + \gamma f_{tr}(\rho_{i,u}^k h_i^k) \tag{7.12}$$

where γ is the weight between the two costs. The first three parts constitute
the energy consumption for duration t and the last one is network traffic which
here denotes the network usage.

7.5.1.2 Formulation

In the following, we formulate the CCS-SS problem to minimize the combination cost function. The optimization problem is shown as follows:

$$\begin{aligned}
&\min \ f\left(h_i^k, \rho_{i,u}^k\right) \\
&s.t. \ C1: \quad \sum_i \rho_{i,u}^k h_i^k = 1 \\
&\quad\quad C2: \quad \sum_k h_i^k o_k^A \leq c_i^A, \forall i \in \mathbf{S}_A \\
&\quad\quad C3: \quad \sum_k h_i^k c_k^B \leq c_i^B, \forall i \in \mathbf{S}_B \\
&\quad\quad C4: \quad x_l = \sum_{k,i,u} m_u^k \rho_{i,u}^k h_i^k r_l^{i,u} \leq c_l, \\
&\quad\quad\quad\quad\quad \forall i \in \mathbf{S} \cup \{o\} \\
&\quad\quad C5: \quad h_i^k \in \{0, 1\}, \rho_{i,u}^k \in [0, 1]
\end{aligned} \tag{7.13}$$

Constraint (1) specifies that user u's demand rate for service k can be served by several servers simultaneously depending on caching/computing strategy and server selection strategy. Constraints (2) and (3) specify the caching/computing resource capacity limits on in-network server nodes. Data rates are subject to a link capacity constraint (4).

Problem (7.13) is difficult to solve based on the following observations:

■ Both the objective function and feasible set of (7.13) are not convex due to the binary variables h_i^k and the product relationship between h_i^k and $\rho_{i,u}^k$.

■ The size of the problem is very large. For instance, if there are F service elements and N network nodes, the number of variables h_i^k is N^F. In future networks, the number of service elements and network nodes will rise significantly.

As is well known, an MINLP problem is expected to be NP-hard [16], and a variety of algorithms are capable of solving instances with hundreds or even thousands of variables. However, it is very challenging to find the global optimum resource allocation in our problem, since the number of variables and the constraints grow exponentially. How to efficiently solve it is left for future research. Instead, in the next subsection, we propose an NLP formulation from a different view which can be solved analytically.

7.5.2 Caching/computing capacity allocation

In this section, we focus on the optimal caching/computing capacity allocated for each service k, but not the optimal cache location. We denote n_{k_1} as the number of content k_1's copies and m_{k_2} as the number of computation k_2's dedicated VM copies. For simplicity, we remove the integer constraint of n_{k_1} and m_{k_2}. In a network with N nodes, if service k can be provided by n server nodes, $N - n$ nodes have to access the service via one or more hops in a steady state. For the irregular and asymmetric network, the authors in [14] take a semi-analytical approach in deriving the average hop distance to the servers, d, as a power-law function of n:

$$d(n) = A(N/n)^\alpha \tag{7.14}$$

Thus, we have

$$d_{k_1}^A = A(N/n_{k_1})^\alpha, d_{k_2}^B = A(N/m_{k_2})^\alpha \tag{7.15}$$

where $d_{k_1}^A$ and $d_{k_2}^B$ are the average hop distance to content k_1's copies and computation k_2's dedicated VM copies, respectively.

We assume the average end-to-end latency is proportional to the average hop distance and the scaling factor is η. Then the network traffic for duration

t is

$$T_{total} = \sum_{k_1 \in K_A} T_{k_1}^A + \sum_{k_2 \in K_B} T_{k_2}^B$$

$$= \eta \left(\sum_{k_1 \in K_A} \lambda_{k_1}^A o_{k_1}^A d_{k_1}^A + \sum_{k_2 \in K_B} \lambda_{k_2}^B o_{k_2}^B d_{k_2}^B \right) \qquad (7.16)$$

According to the energy model in section 7.4.5, the total energy consumption for duration t is as follows:

$$E_{total} = E_{total}^A + E_{total}^B$$

$$= \sum_{k_1 \in \mathbf{K}_A} \left(E_{ca,k_1}^A (n_{k_1}) + E_{tr}^{A,k_1} (n_{k_1}) \right)$$

$$+ \sum_{k_2 \in \mathbf{K}_B} \left(E_{static}^{B,k_1} (m_{k_2}) + E_{active}^{B,k_2} + E_{tr}^{B,k_2} (m_{k_2}) \right) \qquad (7.17)$$

Thus, the caching/computing capacity allocation problem can be formulated as:

$$\begin{aligned} \min \quad & E_{total} + \gamma T_{total} \\ s.t. \quad & 1 \le n_{k_1} \le N \\ & 1 \le m_{k_2} \le N \end{aligned} \qquad (7.18)$$

We ignore the server capacity constraint by assuming sufficiently large server capacities, since we are more interested in the impact of caching/computing capacity allocation on total network cost than the decision of load balancing among congested servers.

By using the Lagrangian dual method, the optimal n_{k_1} and m_{k_2}, which are denoted $n_{k_1}^*$ and $m_{k_2}^*$, can be derived as:

$$\begin{aligned} n_{k_1}^* &= \max[1, \min[n_{k_1}^o, N]] \\ m_{k_2}^* &= \max[1, \min[m_{k_2}^o, N]] \end{aligned} \qquad (7.19)$$

where $n_{k_1}^o$ and $m_{k_2}^o$ are given by

$$n_{k_1}^o = \left[\frac{A\lambda_{k_1}^A \alpha \left(p_{tr,link} + p_{tr,node} + \gamma\eta \right)}{p_{ca}^A t} \right]^{\frac{1}{\alpha+1}} N^{\frac{\alpha}{\alpha+1}} \qquad (7.20)$$

$$m_{k_2}^o = \left[\frac{A\lambda_{k_2}^A o_{k_2}^B \alpha \left(p_{tr,link} + p_{tr,node} + \gamma\eta \right)}{p_{static}^B t} \right]^{\frac{1}{\alpha+1}} N^{\frac{\alpha}{\alpha+1}} \qquad (7.21)$$

7.5.3 The exhaustive-search algorithm

The CCS-SS problem (7.13) can be denoted by minimizing $f(\mathbf{H}, \mathbf{P}|\mathbf{M}, \mathbf{R})$ with the caching, computing, link constraints. Note that if the caching/computing strategy matrix $\mathbf{B}$ is pre-known, the formulation (7.13) will turn into minimizing $f(\mathbf{P}|\mathbf{M}, \mathbf{R}, \mathbf{H})$ with the link constraints, which is a linear optimization problem. In section 7.5.2, we derived the optimal deployment copies $n_{k_1}^*$ for each content k_1, and $m_{k_2}^*$ for computation service k_2 without regard to copy locations in the network. We assume that the element numbers of $\mathbf{K}_A$, $\mathbf{K}_B$ are F_1, F_2. Thus, the number of all possible combination subsets of caching/computing strategy (copy locations) Q are

$$Q = \prod_{k_1 \in \mathbf{K}_A} C_N^{n_{k_1}} \prod_{k_2 \in \mathbf{K}_B} C_N^{m_{k_2}} \tag{7.22}$$

which is significantly reduced in contrast with the original number $2^{N(F_1+F_2)}$. In the following, we propose an exhaustive-search algorithm to find optimal resource allocation solutions for each service.

We denote the set of all possible combination subsets as

$$\mathbf{\Phi} = \{\mathbf{H}_1, \mathbf{H}_2, \ldots, \mathbf{H}_Q\} \tag{7.23}$$

Then, the resource allocation process is described as follows. After the controller selects a caching/computing strategy $\mathbf{H}_q \in \mathbf{\Phi}$, it minimizes $f(\mathbf{P}|\mathbf{M}, \mathbf{R}, \mathbf{H})$ with the link constraints, finding the optimal server selection $\mathbf{P}$ for service requests to minimize the combination cost function f. By exhaustive searching, we choose the optimal caching/computing strategy $\mathbf{H}^*$:

$$\mathbf{H}^* = \arg\min_{\mathbf{H}_q \in \mathbf{\Phi}} f(\mathbf{P}|\mathbf{M}, \mathbf{R}, \mathbf{H}_q) \tag{7.24}$$

and the corresponding $\mathbf{P}^*$ and f^*.

7.6 Simulation results and discussion

To evaluate the performance of our proposed SD-NCC framework, we conduct simulations based on a hypothetical United States backbone network US64 [14], which is a representation of the topological characteristics of commercial networks. We evaluate energy consumption and network usage by comparing our proposed scheme with the traditional network without in-network caching/computing. In addition, we also show the impact of the deployment numbers on system performance.

In the simulation, the parameters of the network equipment are as in[15]. The energy density of a router is on the order of $p_{tr,node} = 2 \times 10^{-8} J/bit$. The energy density of a link is $p_{tr,link} = 0.15 \times 10^{-8} J/bit$. The power density of caching is $p_{ca}^A = 0.25 \times 10^{-8} W/bit$. The power density of computation VM in a static state is $p_{static}^B = 50W$. We assume both the content and computation traffic demands per second are $1GB$.

7.6.1 Network usage cost

In Figure 7.6, we plot the average network traffic per second as a function of the caching/computing nodes' numbers. Compared with the traditional

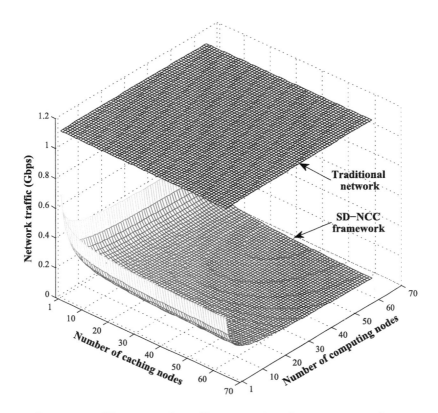

Figure 7.6: The network traffic comparison between two schemes.

network without in-network caching/computing capability, our proposed SD-NCC architecture significantly reduces the traffic traversing the network. The performance advantage is attributed to the fact that the large amount of content and computing requests are served on in-network caching/computing nodes. We also can observe that the traffic reduced ratio declines slowly with the decreasing of deployment numbers of caching/computing nodes.

7.6.2 Energy consumption cost

Figure 7.7 shows the average energy consumption per second as a function of the caching/computing nodes' numbers. With the appropriate deployment numbers of caching/computing nodes, SD-NCC outperforms traditional networks without in-network caching/computing in terms of energy consumption. Besides, with the increasing deployment numbers, the energy consumption decreases at the first stage and increases at the second stage. One can minimize the energy consumption by deploying the optimal numbers of caching/computing nodes. This is due to the fact that caching/computing

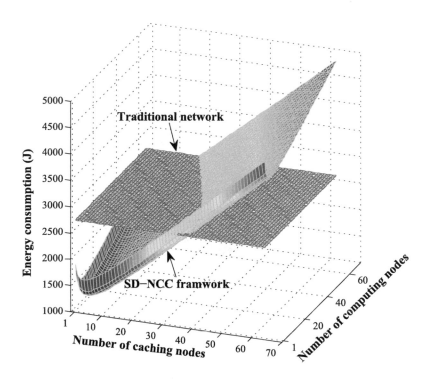

Figure 7.7: The energy consumption comparison between the two schemes.

energy consumption is proportional to the deployment numbers but transmission energy consumption is inversely proportional to the deployment numbers.

7.6.3 Optimal deployment numbers

In section 7.5.2, we derive the optimal deployment numbers on minimizing the combination costs of energy consumption and network usage. Figure 7.8 shows that under the fixed traffic demand, the optimal deployment number of caching copies for each content is proportional to the service popularity but the optimal deployment number of computation copies for each computation is independent of it. Because of the larger number of the same content requests (from different users), less energy is spent on caching (less content need to be cached). In contrast, the energy consumption of computing is fixed, since computation tasks are more complex and the computation result is difficult to be used for other computation requests, even for the same computation requests from different users.

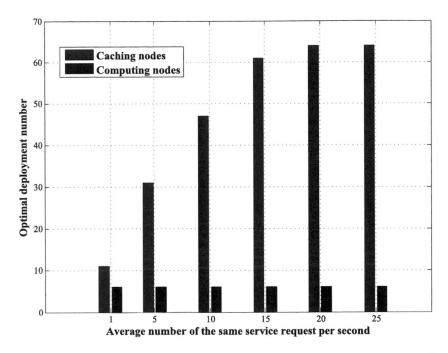

Figure 7.8: The impact of service popularity on optimal deployment number of service copies.

7.7 Open research issues

Despite the potential vision of software-defined networking, caching and computing, many significant research challenges remain to be tackled by future research efforts. In this section, we discuss some of these open research issues.

7.7.1 Scalable SD-NCC controller design

In the proposed framework, the control plane has a logically centralized controller to control networking, caching and computing operations by populating the flow tables of the data plane elements. According to the flow table rule, the forwarding and processing operations of packets in the data plane elements are based on the service names of packets, which include the address of communication endpoints, object names (e.g., content and computation), commands, etc. It is necessary to design a scalable and effective name space system. To address this issue, SD-NCC can learn from the experience of the existing application name space in today's Internet (e.g., DNS), which is well established with name space allocation strategies and governance systems. The performance of the controller could be degraded due to the frequent and rapid flow table update requests as well as caching and computing. Accordingly, it is desirable to design a controller that can handle larger flow tables. Further research is needed for the design of a scalable controller in SD-NCC. We discuss some possible solutions as follows. To solve the scalability issue of the SD-NCC controller, a cluster of distributed controllers can be used. The control plane elements can be physically distributed, but the centralized network-wide view can still be maintained. In addition, parallel computation in multicore systems and improved I/O can be employed to alleviate the scalability challenge of the SD-NCC controller.

7.7.2   Local autonomy in the SD-NCC data plane

The physical devices in the data plane are embedded with intelligence to take local autonomous decisions for networking, caching and computing. By delegating some work to the data plane devices, SD-NCC enables a more scalable and efficient system by reducing the control plane overhead and increasing the flexibility to cope with sudden changes in the network. Note that local strategies in data plane devices are preset by the controller through southbound communications protocols. A modified OpenFlow can be used as the southbound communications protocol between the control plane and the data plane. Moreover, it is necessary to investigate how much local autonomy of the data plane is given by the controller. Here we provide some proposals as follows. In the networking strategy, for instance, some of the routing tasks can be delegated to the data plane to bring positive impacts on routing protocols' design and operation, since local forwarding strategies are more resilient to detect and recover from failures. For the caching strategy, we can release a part of the caching spaces for local decisions to achieve flexible data delivery. The local caching algorithm can be dynamically adjusted according to requirements. Effective caching algorithms in SD-NCC need further research. Likewise, for the computing strategy, computing capability can be partly utilized by the controller, and some computing services no longer need to be reported to the controller, thus achieving a lower latency. It is interesting and challenging to explore the potential of the combination of centralized computing service and local computing service.

7.7.3   Networking/caching/computing resource allocation strategies

SD-NCC provides performance improvement through global coordination of networking, caching and computing resources. The numbers and locations of networking, caching and computing nodes will significantly impact system performance, such as energy consumption, network usage and average service latency. There will be different optimal resource placement strategies corresponding to different system objectives. For example, if the objective is to minimize network energy consumption, an optimal resource placement strategy could be obtained through the trade-off between transport energy and caching/computing energy consumption. Based on the fixed numbers and locations of networking, caching and computing nodes, the caching/computing strategies on nodes are also important for system performance. With a centralized controller, SD-NCC naturally supports cooperation between different caching nodes. Compared with ICN, in which the caching algorithms have been widely studied in recent years, more research on caching cooperation in SD-NCC is needed. In addition, many computing technologies in cloud computing can be extended to the SD-NCC, such as computing resource isolation, distributed computing processing and loading balance. Compared with traditional paradigms, one of the challenges raised by SD-NCC is the complexity of resource allocation. Since SD-NCC needs to handle a large number of networking, caching and computing resources, how to design an effective algorithm with low complexity is challenging, and needs further research.

7.8 Conclusions

In this chapter, we reviewed recent advances in networking, caching and computing. Then, we proposed a framework called SD-NCC to integrate networking, caching and computing in a systematic framework to improve system performance. The design details of its key components of data, control and management planes, as well as the interaction workflow among network nodes were described. Besides, we presented its system model and formulated the joint caching, computing and bandwidth resource allocation problem to minimize the energy cost and network usage cost. Specifically, we derived the optimal deployment number of service copies and used an algorithm to find the near-optimal caching/computing strategy and server selection. Simulation showed that compared with the traditional network, our proposed SD-NCC framework significantly reduces the traffic traversing the network and has a performance advantage in energy consumption. In addition, we discussed open research challenges, including scalable SD-NCC controller design, local autonomy in the SD-NCC data plane, and resource allocation strategies. Future work is in progress to address these challenges.

References

[1] D. Kreutz, F. Ramos, and P. Esteves Verissimo, "Software-defined networking: A comprehensive survey," *Proceedings of the IEEE*, vol. 103, no. 1, pp. 14–76, Jan. 2015.

[2] G. Xylomenos, C. N. Ververidis, and V. A. Siris, "A survey of information-centric networking research," *IEEE Commun. Surveys Tutorials*, vol. 16, no. 2, pp. 1024–1049, 2014.

[3] C. Liang, F. R. Yu, and X. Zhang, "Information-centric network function virtualization over 5G mobile wireless networks," *IEEE Network*, vol. 29, no. 3, pp. 68–74, May 2015.

[4] M. Armbrust, A. Fox, R. Griffith, and A. D. Joseph, "A view of cloud computing," *Commun. ACM*, vol. 53, no. 4, pp. 50–58, Apr. 2010.

[5] M. Zhanikeev, "A cloud visitation platform to facilitate cloud federation and fog computing," *Computer*, vol. 48, no. 5, pp. 80–83, May 2015.

[6] S. Sarkar, S. Chatterjee, and S. Misra, "Assessment of the suitability of fog computing in the context of Internet of things," *IEEE Trans. Cloud Computing*, vol. PP, no. 99, 2015.

[7] ETSI, "Mobile-edge computing introductory technical white paper," ETSI White Paper, Sept. 2014.

[8] N. McKeown, T. Anderson, H. Balakrishnan, G. Parulkar, L. Peterson, J. Rexford, S. Shenker, and J. Turner, "Openflow: Enabling innovation in campus networks," *SIGCOMM Comput. Commun. Rev.*, vol. 38, no. 2, pp. 69–74, Mar. 2008.

[9] J. Liu, S. Zhang, N. Kato, H. Ujikawa, and K. Suzuki, "Device-to-device communications for enhancing quality of experience in software-defined multi-tier LTE-A networks," *IEEE Network*, vol. 29, no. 4, pp. 46–52, Jul. 2015.

[10] L. Cui, F. R. Yu, and Q. Yan, "When big data meets software-defined networking (SDN): SDN for big data and big data for SDN," *IEEE Network*, vol. 30, no. 1, pp. 58–65, Jan. 2016.

[11] J. Li, K. Ota, and M. Dong, "Network virtualization optimization in software-defined vehicular ad-hoc networks," in *Proc. IEEE 84th Vehicular Technology Conference (VTC2016-Fall)*, Sep. 2016.

[12] V. Jacobson, D. K. Smetters, and J. D. Thornton, "Networking named content," in *Proc. the 5th International Conference on Emerging Networking Experiments and Technologies (CoNEXT)'09*, Rome, Italy, Dec. 2009.

[13] Y. Cai, F. R. Yu, and S. Bu, "Cloud computing meets mobile wireless communications in next generation cellular networks," *IEEE Network*, vol. 28, no. 6, pp. 54–59, Nov. 2014.

[14] N. Choi, K. Guan, D. C. Kilper, and G. Atkinson, "In-network caching effect on optimal energy consumption in content centric networking," in *Proc. IEEE ICC'12*, Jun. 2012, pp. 2889–2894.

[15] J. Baliga, R. Ayre, K. Hinton, and R. S. Tucker, "Architectures for energy-efficient IPTV networks," in *Proc. IEEE Optical Fiber Communication Conference (OFC)'09*, Mar. 2009, pp. 1–3.

[16] S. Burer and A. N. Letchford, "Non-convex mixed-integer nonlinear programming: a survey," *Surveys in Operations Research and Management Science*, vol. 17, no. 2, pp. 97–106, 2012.

Chapter 8

Challenges and Broader Perspectives

Despite the promising prospect of the integration of networking, caching and computing, a number of significant research challenges remain to be addressed prior to widespread implementation of integrated networking, caching and computing systems. In this chapter, we present some of these research challenges, followed by a discussion on broader perspectives.

8.1 Challenges

The integrated networking, caching and computing design paradigm raises opportunities as well as challenges in both academia and industry. On the academic side, the classical information theoretic models that anchored numerous breakthroughs in traditional telecommunication technologies cannot be simply applied to integrated networking, caching and computing systems. Similarly on the industrial side, the classical protocols designed for early stages of wireless communication systems are inadequate for leveraging integrated resources on three dimensions.

8.1.1 Stringent latency requirements

One of the foreseeable application scenarios of the integrated networking, caching and computing system is the Internet of Things (IoT) [5]. A lot of industrial control systems, such as oil and gas systems, goods packaging systems, and manufacturing systems, typically require end-to-end latency from sensors to control nodes be bounded within a few milliseconds [6, 7, 8]. Other potential applications of integrated networking, caching and computing systems like virtual reality applications, drone flight control systems, and vehicle-to-roadside communications, often require latency to be under a few tens of milliseconds [5]. These stringent latency requirements fall outside the capability of current systems, and therefore pose a great challenge for integrated networking, caching and computing systems.

8.1.2 Tremendous amount of data against network bandwidth constraints

As an increasing number of mobile devices are connecting to wireless networks, a tremendous and exponentially growing amount of data is being created by them [9]. For instance, the data generated by an autonomous vehicle is estimated to be about one gigabyte per second [10]. The data created by the U.S. smart grid is expected to be 1000 petabytes each year [11], and the U.S. Library of Congress can generate data of 2.4 petabytes per month. By comparison, Google traffics around one petabyte of data a month. Sending all the data to the edge, the fog or the cloud may require extremely high network bandwidth, and in the mean time, cause prohibitively high costs. Furthermore, due to data privacy concerns or regulations, sometimes the data should not be disseminated. Therefore, the central issue here is the design of architectures of integrated systems in which the data can be absorbed (stored) and digested (processed) locally, without long distance propagations.

8.1.3 Uninterruptable services against intermittent connectivity

Some of the devices in the network may suffer from a widely fluctuating quality of wireless channels and intermittent connectivity to edge/fog/cloud servers. Such devices may include vehicles, drones and sometimes mobile terminals of cellular communication systems. However, applications and services like data collection, data analytics and controls need to be available at all times. Therefore, novel system architectures, service time schedulings, caching mechanisms and resource allocation schemes that guarantee consistent availability of services under fluctuating channel conditions and intermittent connectivity to servers are in urgent demand.

8.1.4 Interference of multiple interfaces

In order to provide resources to all types of end users, as well as utilize the caching and computing resources of end users, multiple interfaces (e.g., cellular, WiFi, and WiMAX) should be valid in integrated systems. All users should be able to access the same functionalities through a variety of interfaces.

However, the communication efficiency may be significantly undermined by the coexistence of multiple interfaces due to a paucity of channels. Current technologies that have the potential to enable simultaneous communications via multiple interfaces like SDN, MIMO and cognitive radio in general require multiple access techniques, which are currently not scalable for massive multiple interface wireless networks [12].

8.1.5 Network effectiveness in the face of mobility

In integrated systems, a considerable number of content providers, caching nodes and computing nodes may be mobile devices [13]. In this context, the central issue is how the network can know where to find these caching and computing resources providing nodes. Data request failures and route failures may be increased by these nodes' mobility, and hence deteriorate overall system performance. Therefore, addressing the mobility issue is imperative for the guarantee of network effectiveness.

On the other hand, an integrated system could benefit from the prediction of user mobility. ICN could develop more efficient caching strategies and place the requested content close to users' predicted locations in advance, provided that accurate and timely prediction of user mobility is performed [14]. Moreover, provided with user and group mobility predictions, systems like Mobile Follow-me Cloud (M-FMC) could achieve optimal content migration at different cache levels according to the proximity of users, fulfilling the goals of MEC [14].

Reference [15] takes user mobility into account and proposes a proactive caching approach for named data networking (NDN). In this approach, the content requested but not received by a mobile user before a handover is proactively cached, and is delivered to that user right after the handover. As such, the content delivery ratio can be increased.

In order to analyze and manage mobility in joint microwave (μW)–millimeter wave (mmW) networks, [16] proposes an approach that leverages device-level caching along with dual-mode small base stations. The dual-mode small base stations can utilize both μW and mmW frequencies. Fundamental results on the caching capabilities and average achievable rate of caching are derived, and the impact of caching on average handover failure and energy consumption is analyzed.

However, pertinent research works only offer a preliminary perspective and raise a number of open issues. For instance, one-dimensional mobility models are usually employed in analysis, and one single destination is typically considered, despite the fact that users can pass by multiple destinations in practice. Furthermore, the decision on what services should be migrated and the question of whether group mobility is better than single user mobility have not been adequately addressed [14].

8.1.6 The networking-caching-computing capacity

Determining the theoretical capacity of each dimension of an integrated system is a critical task that remains to be addressed. Since the classical information theoretical model concerning instantaneous rate regions cannot sufficiently address the caching-induced non-causality within the system, it cannot be directly applied to integrated systems. The study in [17] shows that the traditional concept of rate capacity is inadequate to represent the network capability of delivering non-private contents for multiple users. In order to reflect the impact of caching on content delivery, this work proposes a so-called content rate to measure the rate at which the cached content is delivered to users using shared channels. Despite the plethora of studies on the incorporation of computing into wireless networks, the role of computing in the integrated system capacity measurement and calculation remains ambiguous. One category of approaches concerning this matter is networking coding, in which the algebraic operation (i.e., computing) is distinguished from the communication operation [18]. Other approaches like distributed MIMO [19] and base station collaborative transmission [12] do not decouple computing from communication, as channel coding and logic operations are intertwined. Needless to say, a unified capacity analysis in which the networking, caching and computing resources are represented in canonical form (as in Figure 1.1) bears theoretical significance and is therefore in urgent demand.

8.1.7 The networking-caching-computing tradeoffs

The study in [3] demonstrates that the same types of services can be provided by an integrated system utilizing different combinations of networking, caching and computing resources. However, the tradeoff among networking, caching and computing resources for each service type should be analyzed individually against the associated resource constraints and performance measures. In this context, the central issue lies in the optimal tradeoff of resources for each individual case, which is clearly nontrivial to determine.

8.1.8 Security

The new architectures of integrated networking, caching and computing systems present new security challenges. Due to the fact that cloud servers are typically installed in heavily protected facilities selected and controlled by cloud operators, cloud systems are relatively secure against attacks. However, in order to better meet user requirements, fog/edge servers and caching facilities are usually deployed in a distributed manner and in proximity to end users, where users need them to be. Such distributed systems are generally vulnerable to attacks. Moreover, due to limited sizes, these distributed infrastructures may not have sufficient resources to detect threats and protect themselves from attacks.

8.1.9   Convergence and consistency

A considerable number of integrated networking, caching and computing systems are organized in distributed manners, which may cause inconsistency, oscillation and divergence to the global system states, and this may become especially acute in the case of a massive, under-organized mobile system with a virtualized pool of unpredictably shared resources. The convergence and consistency issue is a typical and recurring challenge in distributed systems, and some of the use cases of integrated systems like stream mining and edge analytics have placed exceptional demand on addressing this challenge.

8.1.10   End-to-end architectural tradeoffs

This new paradigm of system design offers new opportunities for creating better tradeoffs between centralized and distributed architectures, between deployment planning and resilience via redundancy, and between what should stay local and what should go global. Moreover, logical integrated system topologies, either dynamically or statically established, over the same underlying physical infrastructures can facilitate the realization of a spectrum of system frameworks from fully centralized to completely distributed.

8.2   Broader perspectives

Since the integrated networking, caching and computing system is currently in its infancy, its development can be influenced by a lot of other technologies. In the mean time, this integrated architecture may impact them as well. Here we briefly discuss these technologies and the potential interactions and benefits when applying these technologies to the integrated system.

8.2.1   Software-defined networking

Software-defined networking (SDN) endows programmability to the networks by introducing a logically software-defined controller and consequently separating the control plane from the data plane [18, 20]. By making network configuration decisions in a centralized controller that has a global view of the network, SDN enables logical centralization of control [5]. Due to its intrinsic virtues, SDN can easily introduce new abstractions in networking, simplify network management, enable dynamic network reconfiguration and facilitate network evolution. Furthermore, SDN can efficiently manage and optimize network resource allocation in response to time-varying network conditions [21]. Due to the effectiveness and efficiency of SDN in managing wireless network systems, it is natural to employ SDN as the control unit of the integrated networking, caching and computing system. However, it is worth noting that the control and resource allocation mechanism of the integrated system is a much broader concept than SDN, and its diversity should go far beyond any single technology.

8.2.2 Network function virtualization

Network function virtualization (NFV) realizes network functions as pure software on commodity and general hardware, which can solve the problem of network ossification, and reduce the CapEx and OpEx as well [22]. NFV is network virtualization on steroids. Instead of controlling network resources, NFV controls network functions such as firewall, load balancer, network address translation, intrusion detection, etc. Network virtualization makes it possible to create multiple virtual networks on top of the same physical network. I contrast, NFV enables creating network functions on top of general-purpose servers instead of the traditional middle boxes.

Recently, there is an increasing trend of integrating NFV with SDN to jointly optimize network functions and resources [22, 23]. SDN and NFV are two closely related technologies. Both SDN and NFV move toward network virtualization and automation; however, they have different concepts, system architectures, and functions that are summarized as follows.

- SDN is a concept of achieving centrally controlled and programmable network architecture to provide better connectivity. In contrast, NFV is a concept of implementing network functions in a software manner.

- SDN aims at providing rapid application deployment and service delivery through abstracting the underlying infrastructure. In contrast, NFV aims at reducing the CapEx, OpEx, space, and power consumption.

- SDN decouples the network control and data forwarding functions, which enables directly programmable network central control. In contrast, NFV decouples the network functions from the dedicated hardware devices, which provisions flexible network functionalities and deployment.

Though there are many differences between them, they are actually highly complementary technologies that can work together. SDN can help NFV in addressing the challenges of dynamic resource management and intelligent function orchestration. Through the implementation of NFV, SDN is able to create a virtual service environment dynamically for a specific type of service chain [22].

With the help of SDN and NFV, an integrated system can realize a dynamic and efficient orchestration of networking, caching and computing resources to improve the overall performance of the entire system.

8.2.3   Wireless network virtualization

Motivated by the success of virtualization in wired networks, the concept of network virtualization has been extended to wireless networks [14]. Wireless network virtualization technology enables the decoupling of wireless network infrastructure from the services that it provides, so that differentiated services can share the same network infrastructure, thereby maximizing its utilization [24]. Wireless network virtualization has the potential to efficiently share and rationally utilize physical network resources, to rapidly respond to dynamic network changes, and to innovatively realize multiple customized services. Furthermore, by isolating part of the network, wireless network virtualization endows the network with the capability of easily migrating to up-to-date technologies while preserving legacy technologies [21]. It should be noted that the sharing of resources enabled by wireless network virtualization is not limited to sharing of networking infrastructure like BSs and routers, but to the extent of sharing of cached content and unoccupied computational resources. That is to say, not only networking but also caching and computing resources can be efficiently shared by end users from different virtual networks through wireless network virtualization. As a consequence, the CapEx and OpEx of wireless access networks and core networks, can be significantly reduced. If network virtualization were applied to an integrated networking, caching and computing system, both the infrastructure and the content within the system could be shared among multiple virtual networks, which exhibits great potential to exploit the three dimensions of resources/capacities to the hilt. However, this in turn increases the complexity of the system, hence raising new requirements on resiliency, consistency, scalability and convergence, as well as resource allocation and management approaches of the entire system.

8.2.4   Big data analytics

Joint consideration of networking, caching and computing will definitely increase the complexity of the system, and raise new requirements on resource allocation and management approaches. In the meantime, an integrated system will inevitably generate a enormous amount of data, such as big signaling data, and big traffic data. We explain more about these two kinds of big data in the following.

Big signaling data

Control messaging plays an important role in networking for mobile users, and the mutual communications between networking, caching and computing functionalities. The control message is also termed signaling data. Signaling data works on the basis of predefined protocols to guarantee wireless communication's security, reliability, regularity and efficiency. In an integrated system of networking, caching and computing, there must be a large amount of signaling data, such as the signaling interaction procedure of the networking-caching framework illustrated in Figure 1.3.

A signaling monitoring system is a significantly important support system for the operation and maintenance of wireless communications, which analyzes rich signaling data and assessment indicators to realize real-time network monitoring, handle customer complaints, troubleshoot the network, block illegal traffic, improve network quality, etc. With the explosive growth of mobile users, and the increasing demand for high-speed data, the amount and types of the signaling data will rise tremendously in an integrated system of networking, caching and computing. Therefore the traditional signaling monitoring systems may be faced with many tough problems, and some insightful information is unable to be discovered. Big data analytics is a recently proposed powerful tool to help analyze big signaling data[25]. For example, in [26], the authors analyze the base station system application part (BSSAP) messages from A interface in a Hadoop platform to identify handovers from 3G to 2G, and the simulation results indicate that the identified 3G coverage holes are in agreement with the drive test conclusions. This example serves as an inspiration for leveraging big data analytics in wireless networks, and it should be promising to incorporate big data analytics into an integrated networking, caching and computing framework to further optimize the system.

Big traffic data

With the development of wireless technologies and the prosperity of social applications, users upload and download lots of video-related content, which results in a vast amount of traffic data that keeps increasing at an amazing rate. Traffic data is continuously generated and carried on in an integrated networking, caching and computing system, and operators have to work hard on balancing the network load and optimizing network utilization.

Traffic monitoring and analyzing is an essential component in network management, which enables the maintenance of smooth network operations, failure detection and prediction, security management, etc. Traditional approaches to monitoring and analyzing traffic data are impotent when faced with the big traffic data. SDN can be combined with big data analytics, which would bring some excellent advantages [1].

The topic of traffic characteristics has attracted a lot of attention, due to the fact that understanding traffic dynamics and usage conditions is of great importance for improving network performance. For example, in [27], the authors use big data analytics to investigate three features of network traffic: network access time, traffic volume, and diurnal patterns. The results reveal some insightful information and provide a reference for operators to take proactive actions for network capacity management and revenue growth. It would be beneficial to introduce big data analytics into an integrated networking, caching, and computing system to facilitate traffic monitoring and network optimization, and acquire insights therein.

8.2.5 Deep reinforcement learning

When we jointly consider the three technologies: networking, caching and computing, system complexity could be very high. Deep reinforcement learning is a recently emerging technology which integrates deep learning with a reinforcement learning algorithm to handle a large amount of input data and obtain the best policy for tough problems. Deep reinforcement learning utilizes a deep Q-network to approximate the Q value function [28]. Google Deepmind adopts this method on some games [28, 29], and gets quite good results. Furthermore, deep reinforcement learning has been explored in various wireless networks[30]. For an integrated system of networking, caching, and computing, deep reinforcement learning can be used as a powerful tool to obtain good resource allocation policies [31].

In summary, research on integrated networking, caching and computing systems is quite broad and a number of challenges lay ahead. Nevertheless, it is in the interest of the wireless community to swiftly address the challenges and go forward. This book attempts to briefly explore the technologies related to the integration of networking, caching and computing at a very preliminary level, and to discuss future research that may benefit the pursuit of this vision. We hope that our discussion and exploration here may open a new avenue for the development of integrated networking, caching and computing systems, and shed light on effective and efficient designs therein.

References

[1] L. Cui, F. R. Yu, and Q. Yan, "When big data meets software-defined networking: SDN for big data and big data for SDN," *IEEE Network*, vol. 30, no. 1, pp. 58–65, 2016.

[2] M. Kiskani and H. Sadjadpour, "Throughput analysis of decentralized coded content caching in cellular networks," *IEEE Trans. Wireless Commun.*, vol. 16, no. 1, pp. 663–672, Jan. 2017.

[3] C. Wang, C. Liang, F. Yu, Q. Chen, and L. Tang, "Computation offloading and resource allocation in wireless cellular networks with mobile edge computing," *IEEE Trans. Wireless Commun.*, vol. 16, no. 8, pp. 4924–4938, May 2017.

[4] C. Wang, C. Liang, F. R. Yu, Q. Chen, and L. Tang, "Joint computation offloading, resource allocation and content caching in cellular networks with mobile edge computing," in *Proc. IEEE ICC'17*, Paris, France, May 2017.

[5] M. Chiang and T. Zhang, "Fog and IoT: An overview of research opportunities," *IEEE Internet of Things J.*, vol. 3, no. 6, pp. 854–864, Dec. 2016.

[6] J. Li, F. R. Yu, G. Deng, C. Luo, Z. Ming, and Q. Yan, "Industrial internet: A survey on the enabling technologies, applications, and challenges," *IEEE Comm. Surveys and Tutorials*, Mar. 2017.

[7] M. Tariq, M. S. A. Latiff, M. Ayaz, Y. Coulibaly, and A. Wahid, "Pressure sensor based reliable (PSBR) routing protocol for underwater acoustic sensor networks," *Ad Hoc and Sensor Wireless Networks*, vol. 32, no. 3-4, pp. 175–196, 2016.

[8] C. Wang and Y. Zhang, "Time-window and Voronoi-partition based aggregation scheduling in multi-sink wireless sensor networks," *Ad Hoc and Sensor Wireless Networks*, vol. 32, no. 3-4, pp. 221–238, 2016.

[9] R. Kelly. (2017) Internet of things data to top 1.6 zettabytes by 2022. [Online]. Available: https://campustechnology.com/articles/2015/04/15/internet-ofthingsdata-to-top-1-6-zettabytes-by-2020.aspx

[10] L. Mearian. (2017) Self-driving cars could create 1gb of data a second. [Online]. Available: http://www.computerworld.com/article/2484219/emergingtechnology/self-driving-cars-could-create-1gb-of-data-a-second.html

[11] N. Cochrane. (2017) Us smart grid to generate 1000 petabytes of data a year. [Online]. Available: http://www.itnews.com.au/news/us-smartgrid-to-generate-1000-petabytes-of-data-a-year-170290ixzz458VaITi6

[12] D. Gesbert, S. Hanly, H. Huang, S. Shitz, O. Simeone, and W. Yu, "Multicell mimo cooperative networks: A new look at interference," *IEEE J. Sel. Areas Commun.*, vol. 28, no. 9, pp. 1380–1408, Dec. 2010.

[13] R. Tourani, S. Misra, and T. Mick, "IC-MCN: An architecture for an information-centric mobile converged network," *IEEE Commun. Mag.*, vol. 54, no. 9, pp. 43–49, Sept. 2016.

[14] A. Gomes, B. Sousa, D. Palma, V. Fonseca, et al., "Edge caching with mobility prediction in virtualized lte mobile networks," *Future Generation Computer Systems*, vol. 70, pp. 148–162, May 2017.

[15] Y. Rao, H. Zhou, D. Gao, H. Luo, and Y. Liu, "Proactive caching for enhancing user-side mobility support in named data networking," in *Proc. International Conference on Innovative Mobile and Internet Services in Ubiquitous Computing*, Taichung, Taiwan, Jul. 2013, pp. 37–42.

[16] O. Semiari, W. Saad, M. Bennis, and B. Maham, "Caching meets millimeter wave communications for enhanced mobility management in 5g networks," *arXiv preprint arXiv:1701.05125* (2017), 2017.

[17] H. Liu, Z. Chen, X. Tian, X. Wang, and M. Tao, "On content-centric wireless delivery networks," *IEEE Wireless Commun.*, vol. 21, no. 6, pp. 118–125, Dec. 2014.

[18] R. Ahlswede, N. Cai, S. Li, and R. Yeung, "Network information flow," *IEEE Trans. Info. Theory*, vol. 46, no. 4, pp. 1204–1216, July 2000.

[19] X. You, D. Wang, B. Sheng, X. Gao, X. Zhao, and M. Chen, "Cooperative distributed antenna systems for mobile communications," *IEEE Wireless Commun.*, vol. 17, no. 3, pp. 35–43, Jun. 2010.

[20] H. Hu, H. Chen, P. Mueller, R. Hu, and Y. Rui, "Software defined wireless networks (SDWN): Part 1 [guest editorial]," *IEEE Commun. Mag.*, vol. 53, no. 11, pp. 108–109, Nov. 2015.

[21] K. Wang, H. Li, F. Yu, and W. Wei, "Virtual resource allocation in software-defined information-centric cellular networks with device-to-device communications and imperfect csi," *IEEE Trans. Veh. Technol.*, vol. 65, no. 12, pp. 10 011–10 021, Dec. 2016.

[22] Y. Li and M. Chen, "Software-defined network function virtualization: A survey," *IEEE Access*, vol. 3, pp. 2542–2553, Dec. 2015.

[23] J. Matias, J. Garay, N. Toledo, J. Unzilla, and E. Jacob, "Toward an SDN-enabled NFV architecture," *IEEE Communications Magazine*, vol. 53, no. 4, pp. 187–193, Apr. 2015.

[24] H. Wen, P. Tiwary, and T. Le-Ngoc, *Wireless Virtualization*. Springer, 2013.

[25] Y. He, F. R. Yu, N. Zhao, H. Yin, H. Yao, and R. C. Qiu, "Big data analytics in mobile cellular networks," *IEEE Access*, vol. 4, pp. 1985–1996, Mar. 2016.

[26] O. F. Celebi et al., "On use of big data for enhancing network coverage analysis," in *Proc. ICT'13*. Casablanca, Morocco, May. 2013, pp. 1–5.

[27] J. Liu, N. Chang, S. Zhang, and Z. Lei, "Recognizing and characterizing dynamics of cellular devices in cellular data network through massive data analysis," *Int. J. Commun. Syst.*, vol. 28, no. 12, pp. 1884–1897, 2015.

[28] V. Mnih, K. Kavukcuoglu, D. Silver, A. A. Rusu, J. Veness, M. G. Bellemare, A. Graves, M. Riedmiller, A. K. Fidjeland, G. Ostrovski et al., "Human-level control through deep reinforcement learning," *Nature*, vol. 518, no. 7540, pp. 529–533, 2015.

[29] D. Silver, A. Huang, C. J. Maddison, A. Guez, L. Sifre, G. van den Driessche, J. Schrittwieser, I. Antonoglou, V. Panneershelvam, M. Lanctot et al., "Mastering the game of Go with deep neural networks and tree search," *Nature*, vol. 529, no. 7587, pp. 484–489, 2016.

[30] Y. He, Z. Zhang, F. R. Yu, N. Zhao, H. Yin, V. C. M. Leung, and Y. Zhang, "Deep reinforcement learning-based optimization for cache-enabled opportunistic interference alignment wireless networks," *IEEE Trans. Veh. Tech.*, 2017, online.

[31] Y. He, N. Zhao, and H. Yin, "Integrated networking, caching and computing for connected vehicles: A deep reinforcement learning approach," *IEEE Trans. Veh. Tech.*, vol. 67, no. 1, pp. 44-55, Jan. 2018.

Index

Page numbers followed by f indicate figure
Page numbers followed by t indicate table